T5-DHA-401

THE INSIDE STORY OF METAL

Metal has always been a mysterious, elusive material. Mostly by trial and error, people learned to master metals, without discovering what is inside them and without knowing what makes them what they arc. This book is intended for the person with a curiosity about metals. Here is everything you need to know about metals: their discovery; metal ores and what they are; the smelting of metals; explanation of metal atoms, crystals, alloys; processes by which metal is cast, forged, etc.; and the relation between metals and energy and resources both today and in the future.

THE INSIDE STORY OF METAL

Norman F. Smith

ILLUSTRATED

Drawings by the author

Julian Messner / New York

Published by Julian Messner, a Simon & Schuster Division of Gulf & Western Corporation. Simon & Schuster Building, 1230 Avenue of the Americas, New York, N.Y. 10020

Printed in the United States of America

Design by Irving Perkins

Library of Congress Cataloging in Publication Data
Smith, Norman F
The inside story of metals.
Includes index.
1. Metals. 2. Metallurgy. I. Title.
TN667.S58 669 77–10768
ISBN 0–671–32860–3 lib. bdg.

This book is dedicated to all the
members of my family who have spent
their careers in the metals industry.

CONTENTS

DRAMATIS PERSONAE

Principals

IRON	The most important metal
ALUMINUM	The light metal
COPPER	The conductivity metal
ZINC	The galvanizing metal
LEAD	The plumbers' metal
TIN	The metal that tins the can
NICKEL	The versatile metal

Supporting Characters

MAGNESIUM, BERYLLIUM	The ultralight metals
TITANIUM	The strong middleweight
CHROMIUM	The stainless metal
TUNGSTEN	The lamp-filament metal
GOLD, SILVER, PLATINUM	The precious trio
IRIDIUM, PALLADIUM, RHODIUM	Their valuable cousins
GERMANIUM	The transistor metal
TANTALUM	The condenser metal
MANGANESE, VANADIUM	The scavenging metals
COBALT, MOLYBDENUM	Some other metals that give new properties to steel
CADMIUM	The weather-resister
OSMIUM	The heaviest metal
LITHIUM	The lightest metal
BARIUM, CALCIUM, POTASSIUM SODIUM	Some reactive metals
MERCURY	The liquid metal
ANTIMONY, ARSENIC, BISMUTH BORON	The half-metals
NIOBIUM, ZIRCONIUM SELENIUM, TELLURIUM	Some new arrivals
CALLIUM, HAFNIUM, INDIUM, RHENIUM	Some rare metals
CERIUM, DYSPROSIUM, ERBIUM, EUROPIUM, GADOLINIUM, HOLMIUM, LANTHANUM, LUTETIUM, NEODYMIUM, PRASEODYMIUM, PROMETHIUM, SAMARIUM, TERBIUM, THULIUM, YTTERBIUM, YTTRIUM	The "rare earth" metals
PLUTONIUM, RADIUM, THORIUM, URANIUM	The radioactive metals
CARBON, SILICON	Two useful "nonmetals"

From *Metals in the Service of Man* by W. Alexander and A. Street

PREFACE

Metal has always been an elusive, mysterious material. Metal was hidden in the rocks and soil that surrounded our ancient ancestors, yet it eluded their grasp for more than two million years. After the lucky accident that gave metal to emerging civilization, metal still remained mysterious and poorly understood.

Largely by trial and error, people learned to master metals, without discovering what is inside them and without knowing what makes them what they are. Only in very recent times has science enabled us to look inside metal to see the wonders that are there and to change the working of metal from a mystery-filled art to a science.

This book is designed to take the reader inside metal and metal technology. Focusing primarily on the science aspects of these topics, the story is intended for the science student or layperson with broad interests, a curiosity about metal, and perhaps some background in school science, chemistry, or physics.

On science topics such as this, the writer's task of setting scope and depth is not an easy one. But it is inescapably his task to sort out those ideas and concepts that he believes will have meaning for his expected audience. He must then transport these ideas from the high, isolated plateau of the technology, across the yawning gulf that separates the technologist from the layperson, through the thorny barriers of language, to the home territory of his readers. The fact that few bridges have been built in the past over these obstacles for the topic of metals has made this task a challenging one.

If I were to attempt to name the books, materials, and people that helped to make this manuscript possible, the list would be long and rambling and not particularly useful. It would include my science and metallurgy professors from years ago, a large number of textbooks on metallurgy, physics, and chemistry, and an assortment of articles, reports, and encyclopedias. It would include several members of my family who spent their careers in the metals industry and a metallurgist who reviewed the manuscript and made many helpful suggestions. The list would include my wife Evelyn, whose perceptive editing and insistence on clarity and coherence have made important contributions to all my books.

NFS

CHAPTER 1

THE DISCOVERY OF METAL

Metal is a very recent discovery. Our ancestors had no metal for their first two and a half million years on earth. They made tools only of stone, wood, and bone. During this time, metal was all around them—in the sand under their feet and in the rock walls of the caves they lived in. It was in the stone from which they made their tools and in the clay from which they fashioned their pottery.

Nearly three-fourths of the elements that make up the earth are metals. If metals are the most abundant elements on earth, why were they not discovered until only a few thousand years ago?

The answer is that metal rarely appears on earth in a pure form—that is, as a free element. Nearly all of the metal in the crust of the earth is tied up with other elements, such as oxygen and sulfur, in chemical compounds called minerals.

Minerals, as found in sand and stone, are quite different from metal and give no hint of the vastly different material hidden within their gritty substance. Few chemical compounds resemble their components. Water, for example, gives no hint that it is a compound of two gases, hydrogen and oxygen.

So early people used stone and other natural materials to make their tools, completely unaware that the metal within the stone, if it could be released, would help make much better tools.

The first use of tools happened so long ago that we can only guess at how it came about. For a long time hunters undoubtedly threw stones to kill birds and used clubs or spears to kill small animals. Perhaps an early hunter cut himself on a sharp stone, and thereby discovered that such a stone could be used to skin a rabbit or to cut up its meat. But a sharp stone picked up and thrown or used without alteration is not really a tool. Some animals, such as monkeys and chimpanzees, can use stones and sticks as weapons. True toolmaking began when our ancestors made *changes* in a rock, stick, or bone they found, to make it more suitable for a certain job. It was the ability to make and use tools that made early people different from the animals around them.

The first tools were probably rocks or pebbles with one side split off to give a sharp edge. They were used for skinning animals and cutting them up for food. Thousands of these rough cutting and chopping tools have been found, some of them made as long as two and a half million years ago. Many of them are so crude that we might not give

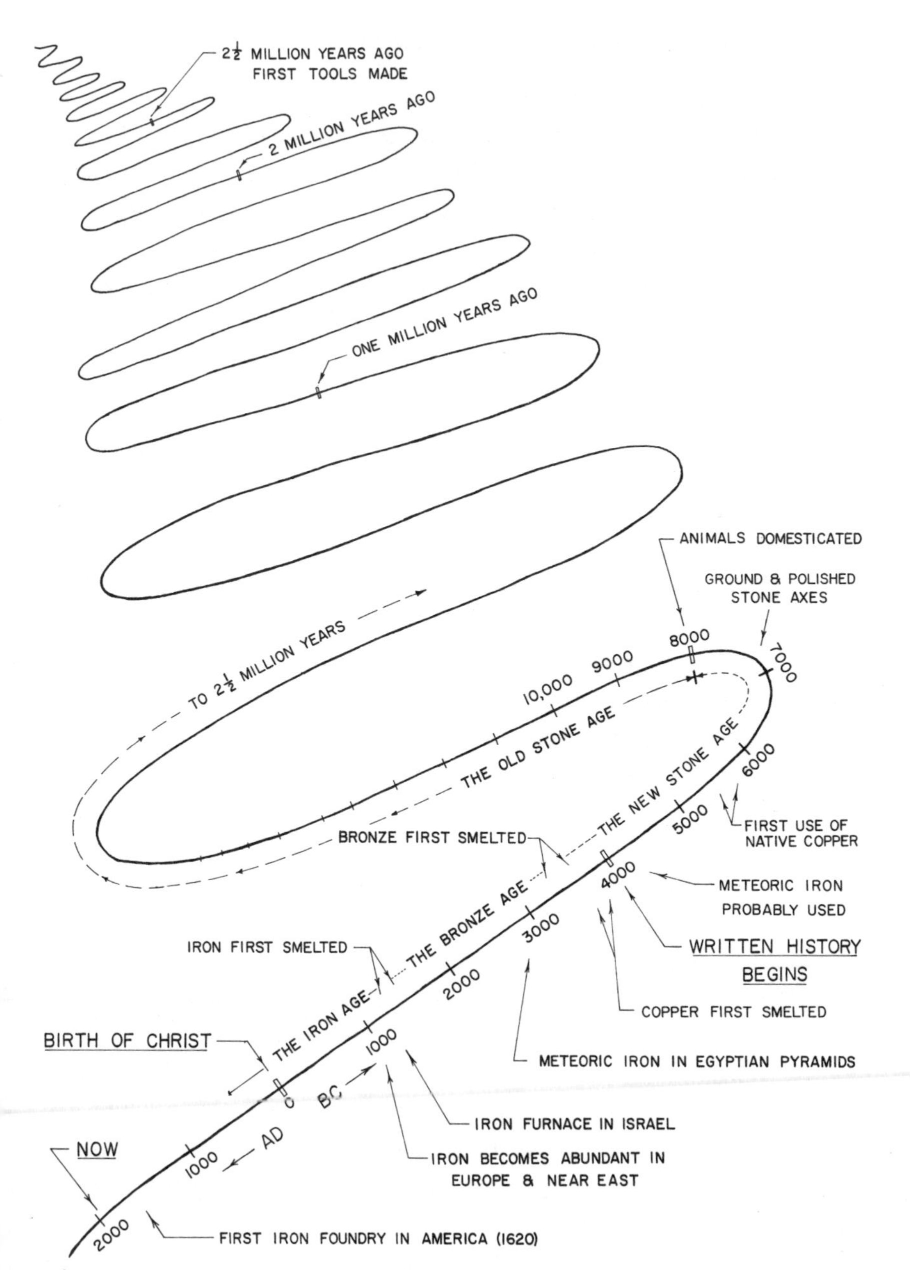

Fig. 1-1. **Important dates in the history of people and metals.**

them a second glance if we saw them in a pile of rocks. But the sharp eyes of the anthropologists—the scientists who study the origins of the human race—can tell that the split edge is not accidental but was made by human hands.

The making of tools by early people improved very, very slowly for more than two million years. We call this period in which stone tools were used the *Old Stone Age,* or the *Paleolithic period* (Fig. 1-1). During this period people learned to choose better stone and to shape that stone into more complicated and more useful tools, such as axes, arrowheads, scrapers, choppers, and knives. They found that the best stone was *flint,* a brittle, grey or smoky-brown stone. Flint is very hard and strong and can be chipped or flaked into tools with very sharp edges. But throughout the Old Stone Age, the stone tools made by our ancestors were still rough and crude-looking (Fig. 1-2).

Then about 10,000 years ago, new ways of shaping and polishing stone were discovered. This was the beginning of a short period called the *New Stone Age,* or the *Neolithic period.* New kinds of stone axes were made in this period, with polished surfaces and sharper blades. These tools were so skillfully made that it took present day scientists many years to learn how to make similar stone blades (Fig. 1-3).

How good were these stone tools? Anthropologists needed the answer to that question in order to learn more about how people lived in that era. They took a stone axehead many thousands of years old from a museum and fitted it with a new wooden handle. When they tried to use this ancient axe to chop down a tree, they found that they had to learn a method of chopping different from that used with a modern axe. But with a little practice they found that they could chop down a 30-cm (12″)-diameter oak tree in as little as

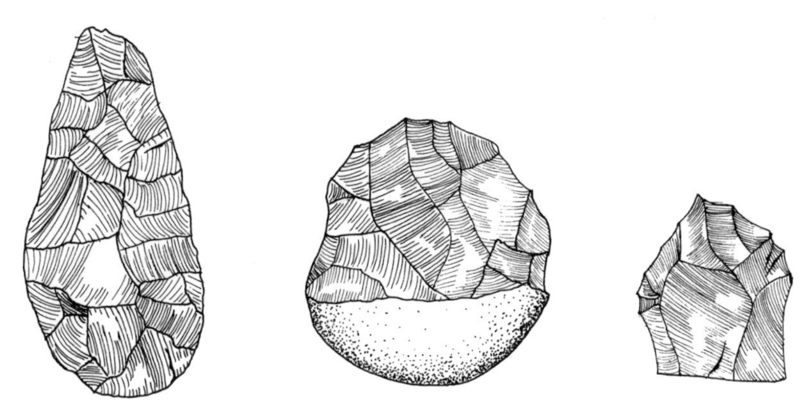

Fig. 1-2. **Tools from the Old Stone (Paleolithic) Age.**

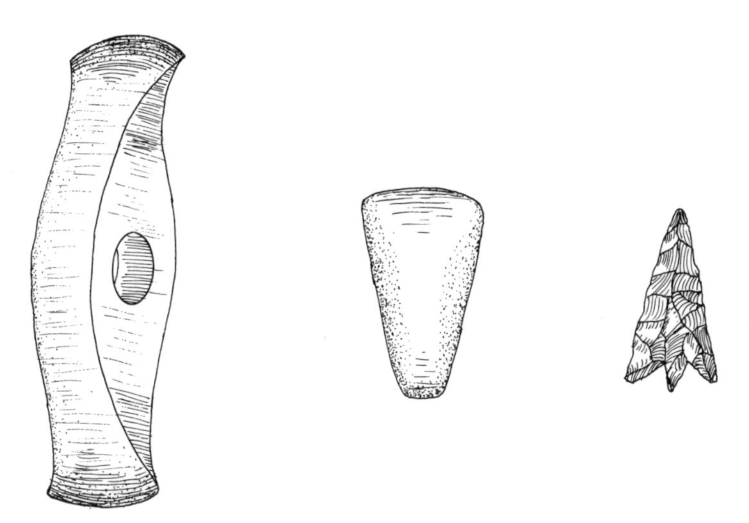

Fig. 1-3. **Tools from the New Stone (Mesolithic) Age.**

30 minutes, and pine trees of up to 60 cm (24″) in diameter in even less time. One stone axe was used to chop down 100 trees before it needed to be sharpened.

From this test and other evidence, anthropologists decided that our ancestors in the Neolithic Age must have been able to use their stone tools to clear away the forest and make the first farms. They probably were also able to shape timber for building homes, to build furniture and household equipment, and to make things that required great skill, such as canoes, paddles, sledges, and skis.

The human race was well on its way toward civilization near the end of the New Stone Age. There werc villages and towns in some parts of the world, although many people still lived in caves or rude huts. Farmers raised grain and vegetables and bred cattle, pigs, goats, and sheep. Spinning and weaving had been invented, and clothing was being made of cloth as well as leather. Fine pottery was being made, some of it beautifully decorated. Furnaces were used for baking pottery at very high temperatures.

Because these Stone Age people left no written history, we are able to learn about how they lived only from the tools and other things found in caves, campsites, graves, and village sites that were used long ago. The first metal things left behind by early people were beads and other small items of gold, copper, and silver. They had found these metals as small lumps or nuggets of more or less pure metal on the ground, among the rocks, or in river beds, just as gold-seekers still do today. Gold and silver are so soft that they could not be used for tools. But these metals are bright and attractive, so they were made into ornaments and jewelry.

Copper was sometimes found in chunks large enough to be useful for making tools. Early metalworkers soon learned that although cooper looks somewhat like gold, it is actually quite different. They found that when they hammered copper into a useful shape (with a stone hammer, of course) the copper became stronger and harder. If shaped into an axe, hammered copper would hold a fairly sharp edge. If it was hammered too much, the metal became brittle and would crack or break easily.

Because copper is neither as strong nor as hard as flint, even the best copper axe could not be made as sharp or as durable as a good stone axe. For this reason, Stone Age people did not throw away their stone tools as soon as metal was discovered. Both kinds of tools were made and used for thousands of years, with metal tools slowly replacing the stone as metal workers learned to produce more and better metal.

Iron was probably discovered somewhat later than copper. Iron objects have been found in Egyptian pyramids that were built about 5,000 years ago. These objects were not made from natural nuggets of iron, because pure iron is seldom found in the earth's crust. Instead, the iron came from meteorites found on the ground. Meteorites are pieces of material, called meteoroids, that come from somewhere in outer space. Most small meteoroids burn up when they strike the earth's atmosphere far above the earth. We see them at night as meteors, or "falling stars." But larger meteoroids —as large as a baseball, for example—may not completely burn up in the atmosphere. The part of a meteoroid that reaches the ground is called a *meteorite*. Most meteorites are made of stone, but some are almost pure metal.

Scientists can always tell whether an ancient piece of iron came from outer space or from the earth. The iron is the same in both, but iron meteorites always contain some of the metal nickel, which is not found in earth iron. Besides helping to identify meteoritic iron, the nickel also helps to keep the iron from rusting away. It is likely that a larger number of ancient items of meteoritic iron have survived than items of earth iron that were made later.

The ancient Sumerians and Egyptians knew, or suspected, that meteorites came from outer space. Perhaps someone saw a meteorite fall and picked it up while it was still hot.

Because copper was the only metal known at that time, the Egyptians called meteorites "black copper from the heavens."

Picking up nuggets of metal from the ground and hammering them into jewelry and small tools was not very difficult and did not represent a very large increase in skill for early people. They had, after all, been picking up rocks, sticks, and pieces of bone to make tools for several million years. But if our ancestors were to have enough metal to make useful tools, they had to discover two new processes: how to melt small pieces of metal to make larger chunks, and how to smelt metal from rocks and minerals. The words *melt* and *smelt* seem much alike, but they describe two very different processes. We *melt* metal by simply applying enough heat to change it from a solid to a liquid. We *smelt* metal when we separate, or refine, a metal from its ore. How were these processes discovered? Which came first? Again, there is no written history to tell us. How the discovery happened can only be guessed from metal objects that have been found and from what we know about the skills of that era.

At one time, archaeologists supposed that early metalworkers figured out first that copper could be melted and then somehow discovered later, probably by accident, how to smelt metal from copper ore. The theory was that a piece of copper ore—that is, rock containing a compound of copper—was dropped into a fire, or perhaps was used as one of the fireplace rocks that held the campfire. After the fire had burned out, according to this theory, the owner noticed some beads of fresh metal in the ashes. This explanation was considered a very likely one for many years. Then someone measured the heat of a campfire and found that an ordinary wood campfire is not hot enough to smelt copper. The first smelting of metal from its ore probably could not have happened accidentally in a campfire.

Where, then, could stone-age people have produced temperatures hot enough to smelt metal? Perhaps you have guessed the answer: in an oven, or "kiln," used to bake pottery! The art of making pottery was well advanced in those days, and pottery was baked—or "fired"—in kilns that could easily be made hotter than a campfire (Fig. 1-4). Also, it happened that raw clay pottery was sometimes smeared with a powder that had been found to give pottery a hard surface after firing. We know that one of the powders often used was a copper ore. On that important occasion, according to this newer theory, the atmosphere in the kiln happened to be just right; the fire was perhaps a little hotter than usual. When the firing was finished and the potter removed his work, he found tiny beads of something on the floor of the kiln. Because he had seen jewelry and other objects made from the same material, he recognized that the beads were metal! Many archaeologists now believe that it was this kind of "accident" in a pottery kiln that produced the discovery that changed the world.

***Fig. 1-4.* This early Greek pottery-kiln from about 700–600 B.C. is probably quite similar to those used 3000 years earlier at the time when metal was discovered. (*New York Public Library Picture Collection*)**

Actually, there was one more fortunate circumstance that helped to bring about the discovery of smelting. Most metals are tightly bound to oxygen in the ore. Heat alone will not release metal from an ore—there must also be what chemists call a "reducing agent" present to get out the oxygen. One of the best reducing agents—one still used throughout the metals industry today—is carbon. Carbon captures the oxygen from the ore and combines with it to form carbon dioxide (CO_2), thereby releasing the pure metal. The fuel that early people used in campfires or kilns was wood. As wood burns, it becomes charcoal, which is almost pure carbon. Thus, every fire that primitive people built in a fireplace or kiln, had in its glowing embers the hot carbon that was needed for smelting metal out of ore. All that was needed was the right kind of ore, a high enough temperature, and someone to recognize the beads of metal on the kiln floor.

The first crude efforts at smelting copper from its ore did not produce pure, shining metal, but rather "sponge copper," a rough porous mixture of metal and rocklike impurities called *slag*. By beating this strange new material with a hammer, early metalworkers found that they could force out the slag and shape the copper into useful tools. Later, when they learned to heat the sponge copper to temperatures high enough to melt it, they found that the slag would float on top of the heavier liquid metal where it could be skimmed off. The liquid copper could then be poured into molds for further use. Here the pottery skills of that time came to the metalworker's aid again. Pottery vessels were available in which metal could be melted. Also, there were pottery molds into which molten copper could be poured. We can imagine how excited our stone-age ancestors must have been when they first succeeded in wresting a stream of bright shiny new metal from dull, worthless rocks.

Most authorities believe that the first smelting of copper took place around 4000 B.C., or perhaps a little later. No one knows exactly where copper was first smelted, but it is believed to have happened in one of the ancient civilizations at the eastern end of the Mediterranean Sea—the area we now call the Middle East. From here the new skill spread slowly through Europe, Asia, and Africa. It did not reach the American continents in ancient times, however, because the land connection between Asia and Alaska had long ago disappeared. The Indians of North and South America did not discover smelting on their own. They were found to have no metal, except for native gold, silver, and copper, when explorers from Europe reached them in the fifteenth and sixteenth centuries.

Although copper proved to be very useful, it had serious limitations, too. Even when hardened by hammering, a copper axe was still too soft to stay sharp for very long. After perhaps 500 years of smelting copper, metalworkers discovered what seemed to be a new kind of copper. We don't know whether the discovery resulted from the accidental smelting of copper ore that happened to contain the metal tin, or from the intentional mixing of known ores of tin and copper. In any event, the new metal was a mixture, or alloy, of copper and a small amount of tin, which today we call *bronze*. Bronze was stronger and harder than copper alone; it was easier to cast, and could be made into axes and knives that would keep sharp edges longer. The simple bronze that early metal workers produced is still an important metal alloy today. This period of history in the Middle East brought an end to the New Stone Age. When copper and bronze began to be produced, human beings entered the *Bronze Age.*

As the methods of smelting copper and bronze improved over the centuries, it was inevitable that other metals would be discovered. About 2,500 to 3,500 years ago, or roughly 2,000 years after bronze was discovered, the first iron was smelted. No one knows how this discovery occurred. It has been suggested that iron ore was accidently used instead of copper ore. Because iron ores are among the most plentiful ores on earth, and because a furnace need not be any hotter for smelting sponge iron than for smelting sponge copper, it is perhaps surprising that iron was not discovered earlier.

As with copper, the first iron that came from the furnace was a spongy mixture of iron and slag. Metalworkers found, probably after some trial and error, that hammering this mass while red hot squeezed out the slag and welded the spongy iron together into a nearly pure form that we now call *wrought iron.* Wrought iron could be easily hammered and bent into tools and weapons.

At first, iron was so soft that it was not as useful as bronze. It did not make very durable weapons. Many a soldier probably bent his sword in battle and had to straighten it over his knee or between two boulders before he could go back into the fray. So metalworkers tried to make their iron harder and stronger. They found that when they heated their iron for a longer time in a hotter fire they produced a much better kind of iron. They didn't know why this happened, but we know now that the red-hot iron absorbed a small amount of carbon from the charcoal. This carbon made a new kind of iron which was much harder, stronger, and tougher than wrought (pure) iron. We call it steel.

Slowly, over several thousand years, better furnaces were

developed. The bellows was invented, which could be used to blow air through a fire to make it burn hotter. Improved metalworking methods were developed to make more and better metal for all sorts of uses. Iron and steel made possible the manufacture of tools for easier land clearing and more productive agriculture; better materials for ships and other forms of transportation. Steel eventually made possible railroads, bridges, skyscrapers, automobiles, and all the products of our modern world. Today we are still in the iron age—or perhaps we should call it the steel age—that began with the first smelting of iron between 3,000 and 3,500 years ago.

Through most of these years since the New Stone Age, the production of metal was not a science but a trial-and-error process. As a new understanding of the nature of metal became available during the past 100 years the science of metals began to emerge. Metallurgy, as this science is now called, has been described as one of the oldest arts and one of the newest sciences.

Today, the science of metallurgy has two branches: *extractive metallurgy,* which deals with ores and all the processes used for extracting metal from these ores, and *physical metallurgy,* which deals with the use of metal, including both the physical characteristics of metals and all the processes used for making and shaping metal into useful things. The next two chapters deal with the work of the extractive metallurgist.

CHAPTER 2

METAL ORES

The metal ores that we mine today did not exist when the earth was formed, four or five billion years ago. At that time, the earth was a hot, largely molten mixture of all kinds of elements. Because of the high temperature of the earth, many metals probably existed then as pure metal—that is, they were not combined chemically with other elements.

The heavier elements, such as iron, lead, and nickel, were pulled by gravity toward the center of the earth. The lighter elements, such as aluminum, silicon, and magnesium, and minerals containing these elements, floated to the top. As a result, scientists believe that the earth today has a core of heavy metal, probably iron and nickel, measuring about 4,000 miles across. As the earth cooled, the outer surface solidified as a crust, made up mostly of lighter elements and their compounds. The crust is between 5 and 30 miles thick.

During this period many heavy metals sank too far beneath the earth's surface to be mined. Some of those metals might have stayed beyond our reach if the earth had cooled down quietly. But instead, the earth rumbled and churned inside. The crust cracked, and volcanoes burst forth from the surface. Molten rock from great depths was forced up through the crust and out over the surface of the earth. Escaping gases helped to push liquid rock that contained metal up into cracks in the crust, where it solidified to become rich veins of ore. Hot gases and water vapor from the *magma* (molten rock that has not reached the surface) seeped up through fissures and broken rock. When these gases cooled, they deposited minerals in the rocks around them. As the earth's crust wrinkled and folded, whole mountains of rock and metal ores were pushed up. Over millions of years the weather wore these mountains down again. Running water carried dissolved or broken pieces of minerals and ores down into low lakes and seas. When the water dried up and the land rose once again, these deposits became the giant ore beds we use today.

While the earth was cooling down, most metals slowly began to combine with other elements such as oxygen and carbon. Some metals, particularly those deeper within the earth, combined with sulfur to form sulfides. Much later, living things appeared on earth and began to make the earth's atmosphere rich in oxygen. The free oxygen in the atmosphere speeded up the combination of the free metals in the earth's crust into oxides. Water trickling down through the rocks over millions of years changed and sometimes moved ore deposits in ways that are not fully understood. Today nearly all metals in the earth's crust are locked up chemically in rocks and minerals as oxides and sulfides from which they must be unlocked if we wish to use them as free metal (Fig. 2-1).

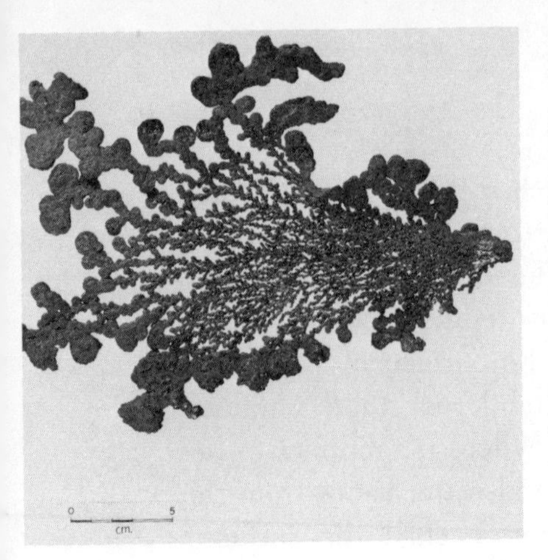

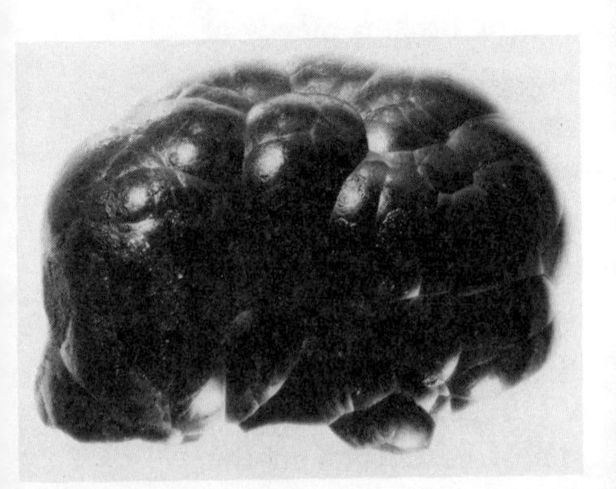

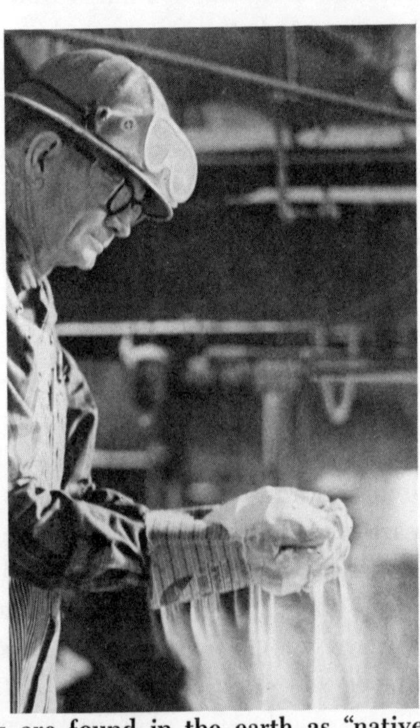

Fig. 2-1. Only a few metals are found in the earth as "native metals"—the rest are locked up in rock-like or soil-like minerals.
top left, Native copper (*Smithsonian Institution*)
top right, Malachite, a copper ore (*Smithsonian Institution*)
bottom left, Hematite, an iron ore (*Smithsonian Institution*)
bottom right, Alumina, an aluminum ore (*Reynolds Metals Co.*)

How tightly a metal is held in an ore depends upon how chemically active the metal is. A very active metal will bind itself very tightly to nonmetals (such as oxygen) to form very stable compounds. An active metal will therefore be more difficult to break loose from its ore than a less active metal. We can see the relative activity of the various metals by consulting a list of elements known as *The Chemical Activity Series* in any chemistry book. Figure 2-2 shows the order of activity for the more common metals according to this series.

Potassium, at the top of the list, is an extremely active metal. It is so strongly attracted to oxygen that it will pull oxygen out of some compounds in order to combine with it.

***Fig. 2-2.* Metals can be arranged according to their chemical activity in a list called the chemical activity series.**

	Metal	*Chemical Symbol*
Very Active ↑	Potassium	K
	Sodium	Na
	Magnesium	Mg
	Aluminum	Al
	Manganese	Mn
	Zinc	Zn
Increasing	Chromium	Cr
Activity of elements	Iron	Fe
and	Nickel	Ni
Stability of compounds	Tin	Sn
	Lead	Pb
	Copper	Cu
	Mercury	Hg
	Silver	Ag
	Platinum	Pt
Inert	Gold	Au

When potassium is exposed to water, for example, even in the form of water vapor, it breaks down the water into hydrogen and oxygen gases and combines with some of the released gases to form a new compound, potassium hydroxide. The heat of the reaction melts the potassium and sets fire to the released hydrogen, creating a violent reaction that may be explosive in force.

Sodium, the next element on this list, is less active than potassium, but will still react strongly with water in much the same way. Magnesium and aluminum combine readily with oxygen in air. In the form of thin ribbon or powder, these metals will burn fiercely when ignited. (Magnesium is the material that burns in photographic flash bulbs.)

All of the metals down to mercury will oxidize in air. However, they will do so less and less easily as we move down the list. Copper will oxidize only at high temperatures, for example, while platinum and gold rarely combine with any elements and are known as "inert" or "noble" metals.

When we examine the chemical activity of different metals in this way, we can see why their ores are all different. Inert metals such as gold and platinum are usually found as pure metal in nature. Silver is often found as pure metal, copper sometimes, and elements above copper rarely. The most active metals toward the top of the list are *never* found as pure metal in the natural state, but are always tightly locked up in compounds.

Because each metal is different from all others, each must be smelted from its ores by a different (though sometimes similar) process. Each ore appears in the earth's crust in different forms and at different depths. Because of these differences, many diverse methods of mining, processing, and smelting ores must be used. A few of the commoner methods

***Fig. 2-3.* A modern head frame for an underground copper mine in the Colorado Rockies. (*Copper Development Association*)**

used in the field of extractive metallurgy will be discussed in this chapter and in Chapter 3.

Ore deposits are dug out of the earth in all sorts of ways. Ores that are deep down in hard rock must be mined by blasting tunnels and holes down into the earth (Fig. 2-3). Some ores can be mined by blasting tunnels into the sides of mountains or hillsides.

Fortunately, the ores of many of our most useful metals lie close to the surface. Much iron ore is dug from shallow pits with power shovels that can lift many tons at one bite. Copper is often mined in a similar way (Fig. 2-4). Aluminum ore is often found near the surface in rocks that can be blasted into pieces and shoveled up.

Ores for some metals are found under water, having been carried by running water and deposited in a low area that later sank beneath the sea. Tin ore is mined in the East Indies by dredging the ocean floor. Chunks, or nodules, of

***Fig. 2-4.* Open-pit copper mine at Twin Buttes, Arizona. (*Copper Development Association*)**

ore rich in manganese oxide have been discovered on the ocean floor—in some places as plentiful as 100,000 tons per square mile. These nodules are believed to have been formed in the process of underwater weathering of volcanic rocks. Such nodules have also been found to contain other metals, such as iron, copper, nickel, cobalt, zinc, and molybdenum. If nodule ores can be harvested from water up to two miles deep, the ocean may one day become a major source of these metals.

Ocean water also contains a great deal of metal, mainly as dissolved compounds. Sodium (in sodium chloride, or salt) is the most plentiful, with magnesium, potassium, and calcium in lesser quantities, in that order. Most other metals are present in tiny proportions—even gold. There are millions of tons of gold in the oceans, but it is so widely distributed that it is not feasible to extract it. The only metal extracted commercially from sea water is magnesium. Although there are ores of magnesium on land, nearly all of the magnesium produced in the United States is taken from seawater.

The ores of various metals differ greatly in composition. The ores most commonly used to produce iron are oxides —that is, compounds of iron and oxygen. They include ferric oxide (chemical formula: Fe_2O_3), ferrous oxide (FeO), and magnetite, or magnetic oxide (Fe_3O_4). The ore usually contains other substances such as oxides of other metals and a considerable amount of worthless, stony material.

Ores of copper that are commonly used contain copper in combination with either oxygen or sulfur. The oxide ores include simple oxides such as cuprite (Cu_2O), tenorite (CuO), and more complicated minerals containing carbon dioxide and water, such as the green crystalline ore malachite [$CuCO_3.Cu(OH)_2$]. The sulfide ores of copper include chalcocite (Cu_2S), covellite (CuS), and others containing additional metals such as iron, antimony, and tin, such as chalcopyrite ($CuFeS_2$). Most copper ores, as mined, contain large amounts of worthless, stony material.

Aluminum minerals are found in nearly all the rocks of the earth. In these minerals, aluminum appears in combination with several elements, including oxygen, phosphorus, sulfur, silicon, and water. Two useful ores of aluminum (loosely known as bauxite) are gibbsite [$Al_2O_3.3H_2O$] and boehmite [$AlO(OH)$], which are complicated minerals containing aluminum oxide.

Some ores are very rich in metal. One hundred pounds of ore might contain more than fifty pounds of metal. Many ores are not that rich. One hundred pounds of some ores may contain only a pound or two of metal, or even less, the rest being rock and soil. Nearly all ores, therefore, must be processed to remove the worthless material and concentrate the ore to be smelted. Usually the ore is processed near the mine, so that the extra weight of the waste material will not have to be transported very far. After the metal-rich part of the ore has been separated from the waste material, it can be shipped to the smelter at a lower cost.

The first step in concentrating ore is usually to crush it into a fine powder. Next, the particles of powder that contain metal ore must be separated from the particles of rock and dirt. Although the tiny specks of crushed ore all look alike, they are always different in some way. The particles of metal ore may be heavier than the rock particles. When the crushed ore is washed in a stream of water, the lightweight particles are swept along by the water, while the heavy particles of metal ore settle to the bottom and are left behind. This process is much like that used by gold prospectors when they "pan" gold from the sand of streams. When the prospector sloshes water and sand around in a pan, the lighter particles rise to the top to be flushed away, while the heavier gold particles settle to the bottom. If the prospector is lucky, he might soon have in the bottom of his pan a small amount of sand and a great deal of gold.

Some kinds of powdered ore that contain metal will stick to air bubbles, while rock particles will not. Oil or some other "foaming agent" is added to the water and air is bubbled through the ore and water. The particles of metal ore that cling to the bubbles can be skimmed off, while the rock particles that sink to the bottom are thrown away (Fig. 2-5).

Some kinds of iron ore are magnetic and are attracted to a magnet just as a piece of iron is. When this ore is run through a magnetic separator, the particles of iron ore cling to the magnet, while the particles of rock drop off and are thrown away.

Fig. 2-5. Concentrating copper ore by the flotation process. (*Kennecott Copper Corporation*)

Grinder for pulverizing copper ore. (*Copper Development Association*)

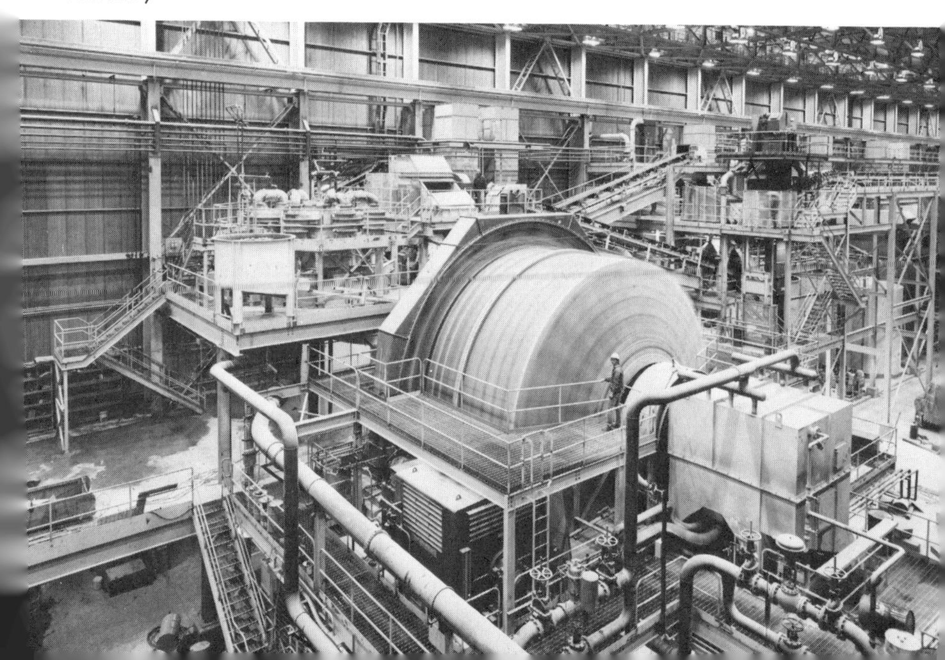

Sometimes several different methods are used, one after another, to be sure that very little usable metal ore is thrown away, and to make sure that as little rock and waste material as possible goes into the processed ore. Although the *concentrated ore* produced by these processes is not entirely pure, the amount of rock and soil left is usually small enough so that it will not interfere very much with the smelting process.

Sometimes the powdered ore is given a special treatment to make it easier to handle. Iron ore, for example, is often combined with clay and baked into little pellets. The pellets will not be carried off by the wind as a powdered ore might. The pellets can be shipped to the smelter in railroad cars or ships and can be easily loaded and unloaded by conveyor belts or other machinery.

Some ores must be concentrated by chemical processes, either because the ore contains only a very small amount of metal, or because the ore must be very pure before it can be smelted. Aluminum ore, for example, contains iron and other impurities that must be removed because they will interfere with the refining process. Aluminum ore, called *bauxite,* is crushed, dissolved in chemicals, and refined in a complex process into a grainy white powder that is almost pure aluminum oxide—Al_2O_3, or alumina. Uranium ore is another that must be refined by a long and complex chemical process before the metal can be smelted out.

Neither the raw ore nor the processed ore looks or acts at all like metal. Iron ore may be small pellets, copper ore a fine powder, and aluminum ore a material that looks like white sand. In order to release the metal from these ores, the metal must be freed from the chemical compounds in which it has been locked up for billions of years. This is done in the process called *smelting.*

CHAPTER 3

SMELTING METALS

The smelting process is somewhat different for each metal. But the goal of each process is always the same: to break atoms of metal loose from the atoms of whatever element is holding them prisoner in the ore. Four main things are needed to smelt most metals: ore, flux, heat, and a reducing agent.

The preparation of ore for smelting was covered in the previous chapter. Although the concentration process removes much of the worthless rock, some of these same impurities and other substances still remain in the processed ore. When the ore is smelted, this waste material may help to spoil the metal by collecting in small globs or spoil the metal in some other way. It is important that this waste material, called *slag*, be separated from the metal.

Separation of the slag is the task of the *flux*. Two common fluxes are silica, which is found in beach sand, and ground-up limestone. The flux makes the globs of slag clump together in a lightweight layer that will float on top of the metal. It may also make the crusty slag more runny so that it can be drained off the top of the molten metal more easily.

Both the heat and the reducing agent are often supplied by the same material—the fuel. For early iron smelting, charcoal was the fuel that did both jobs. Charcoal was made by partially burning wood or by heating wood in the absence of air. Either way, the water and gases in the wood are driven out, leaving only black carbon which is pure enough for use in smelting. (Charcoal burns with a clean, even heat, which also makes it useful for cooking on outdoor barbecue grills.)

Iron smelting in Europe was consuming a huge amount of charcoal at about the time the American Colonies were being settled. The demand was so great that many of England's forests were cut down to make charcoal. A shortage of wood forced England to decrease its iron smelting for a time and buy much of its iron from the Scandinavian countries which still had plenty of wood.

Metalworkers tried to use coal (which was plentiful in England) in their furnaces. They found that the gases and impurities in the coal, such as sulfur and hydrogen, interfered with the smelting process. Around 1708, an Englishman named Abraham Darby tried to make charcoal out of coke. He found that when the gases and impurities were driven out of coal, there was left a dull, porous material that no longer looked like coal. This material was found to be mostly carbon, chemically quite similar to wood charcoal, and pure enough to use in smelting metals. It is called *coke.* Darby became the first (1735) to successfully substitute coke for wood charcoal. Soon, the British began to use the new coke in the manufacture of iron and steel. The American Colonies did not switch to coke until much later. By the time they stopped using charcoal, a great deal of timber in the eastern part of the continent had been cut down and converted to charcoal for use in American iron furnaces.

Huge quantities of coal are now used throughout the world to make coke for the metals industries. Many iron works have their own ovens for converting coal into coke. The gases roasted from (that is, forced out of) the coal are burned as fuel in the plant, while the coke goes to the smelting furnaces.

Smelting processes are described below for three of our most useful metals. These processes illustrate some of the most common ways in which ore, heat, reducing agent, and flux are used in the metals industry. Some steps have been left out or simplified, because we are interested here primarily in *what* is done to the ore as it is refined into metal, rather than in the detailed technology of just *how* it is done. We will focus, in other words, on the science rather than on the technology.

Smelting Iron

Iron is refined from its ore today in much the same way as it was smelted several thousand years ago. But the small, crude furnaces in use then produced only a few pounds of metal per day. Today the demand for iron and steel is so great that huge furnaces have been built that can smelt as much as fourteen million pounds of iron in a single day. Much of the iron produced is processed, often in the same plant, into various kinds of steel. The iron and steel industry in the United States is a mass-production industry that puts out around 100 billion pounds of metal per year. A modern plant gobbles boatloads of ore and trainloads of fuel, pour-

Fig. 3-1. **A blast furnace is used to smelt iron from ore. Note piles of ore and limestone at left. (*American Iron & Steel Institute*)**

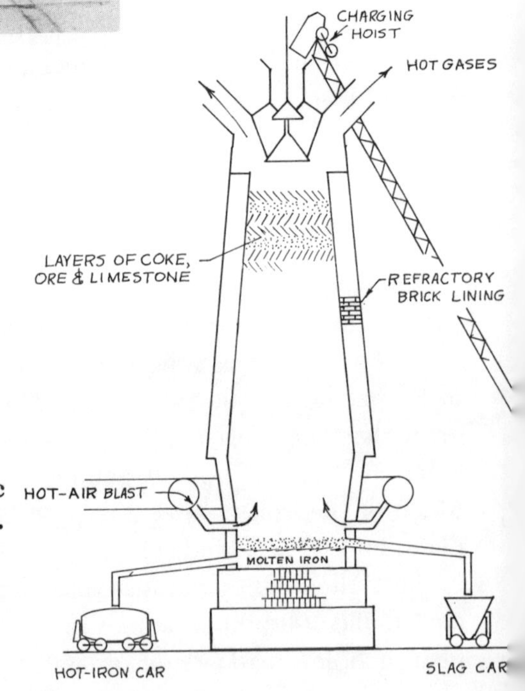

Fig. 3-1a. **Schematic view of a blast furnace.**

ing out trainloads of iron and steel products of all kinds.

The smelting of iron ore is done in a blast furnace which may be nearly ten stories high (Fig. 3-1). The steel shell of the furnace is lined with firebrick that will resist very high temperatures. The iron ore is fed, along with coke and flux (usually ground-up limestone), into the top of the furnace. Around the lower part of the furnace are water-cooled nozzles through which very hot air is blown. As the hot air roars up through the furnace, it is heated still hotter by the burning coke. The glowing carbon in the coke breaks the oxygen loose from the iron-oxide ore and combines with it to form carbon dioxide, which escapes into the atmosphere (Fig. 3-2). The limestone flux combines with the rock and other impurities and carries them off as a frothy mixture of partially molten rock—the slag. The iron, now free from the slag and from the oxygen that held it in the form of ore, trickles down through the furnace as molten iron and is drawn from the bottom of the furnace into molds or ladles.

Once a blast furnace is placed in operation, it runs day and night and shuts down only for repairs. Raw materials

***Fig.* 3-2. The smelting process for iron.**

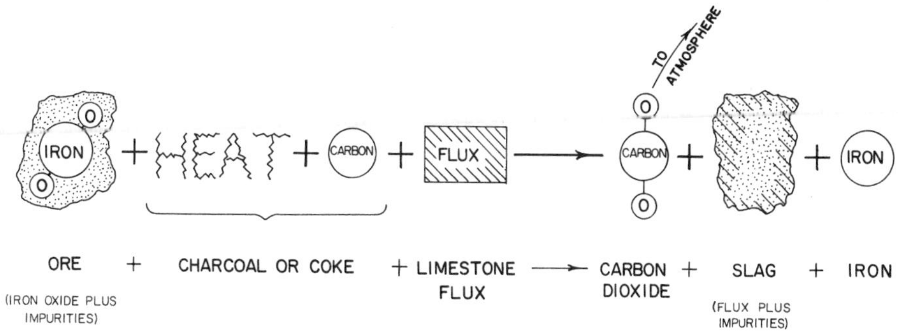

TO ATMOSPHERE
O
IRON
O
+ HEAT + CARBON + FLUX ⟶ CARBON + + IRON
O
O

ORE + CHARCOAL OR COKE + LIMESTONE FLUX ⟶ CARBON DIOXIDE + SLAG + IRON

(IRON OXIDE PLUS IMPURITIES)

(FLUX PLUS IMPURITIES)

pour into the top of the furnace, and a stream of molten iron is drawn periodically from the bottom. The chemical equation for a typical smelter reaction is:

$$2Fe_2O_3 + 3C \longrightarrow 4Fe + 3CO_2$$

$2Fe_2O_3$ +	$3C$	$\longrightarrow$	$4Fe$ +	$3CO_2$
(iron oxide)	(Carbon)		(iron)	(carbon-dioxide)

The molten iron drawn from a blast furnace is sometimes poured into molds. Originally these molds were merely hollowed-out troughs in a bed of sand. The iron ran into a long trough and then overflowed into many smaller troughs which branched off at right angles. Because this arrangement looked a little like a litter of pigs nested against the mother sow, the long piece became known as the "sow," while the smaller chunks were called "pigs." In later years, when iron was poured into different kinds of molds, there were no longer any "sows" and "pigs." But the names stuck. Raw iron from a blast furnace is known today as "pig iron," whether it is poured into molds or taken directly as molten metal to other furnaces for processing into steel.

Pig iron usually has absorbed so much carbon from the burning coke that the iron is weak and brittle. Pure iron can be produced by burning the carbon out of the molten pig iron, using air or oxygen. In a *puddling furnace,* raw iron ore (which, you will recall, is iron oxide) is stirred into the molten pig iron. Air or oxygen is also added to the atmosphere in the furnace. The molten iron roasts oxygen out of the iron ore, which bubbles up through the iron, combining with the carbon, carrying it away as carbon monoxide and carbon dioxide. At the same time, other impurities are also removed by the oxygen, leaving almost pure iron.

Pure iron is called *wrought iron* and is much like the iron produced by very early metalworkers. (The temperatures used by early metalworkers were not high enough to cause the iron to absorb carbon; consequently, the metal they puddled out of their crude furnaces was more like wrought iron than pig iron.) It is soft and easy to pound into various shapes, but not very strong. It is used today mostly for ornamental purposes such as lighting fixtures, hardware, furniture, and railings. The processing of iron into cast iron and various kinds of steel will be described in Chapter 7.

Smelting Copper

Copper was the first metal used by people for tools and weapons and is still one of our most important metals. It is used today mostly for electrical wire and equipment, plumbing materials, and for hardware that will resist corrosion.

Copper can be produced by several methods, depending upon whether the copper in the ore is combined with oxygen (for example, Cu_2O) or sulfur (for example, Cu_2S). For oxide ores, which are used to provide only a small percentage of the world's copper, the refining can be done in a blast furnace. This process is similar in principle to that described earlier for iron; the carbon in the coke captures the oxygen from the ore, leaving metallic copper.

For the more common sulfide ores, the refining process has several steps. First, the ore may be roasted to drive off part of the free sulfur and to oxidize part of the iron sulfide that is present as an impurity. At this stage, the ore is a mixture of copper sulfide, iron sulfide, and iron oxide. The ore is then melted in a furnace, along with flux. Some of the impurities, including iron oxide, are slagged off, leaving a combination of iron sulfide and copper sulfide known as *matte.* The matte is drawn from the furnace and taken to a converter, which is a cylindrical vessel with ports at the bottom for admitting air. When air is blown up through the molten metal, through these ports, the iron sulfide is burned out of the matte. Then the copper oxide reacts with the copper sulfide to produce metallic copper and sulfur dioxide gas. A typical chemical reaction is:

$2Cu_2O$	+	Cu_2S	$\longrightarrow$	$6Cu$	+	SO_2
(copper oxide)		(copper sulfide)		(copper)		(sulfur dioxide gas)

Flux carries away the impurities as slag.

The copper now is called *blister copper.* It usually contains a significant amount of impurities. Often the copper is given another fire-refining operation before being used in a final product. If the copper is intended for electrical use, it may be given a final refining by *electrolysis.* A slab of blister copper is placed, along with another copper electrode, in a tank of water in which copper sulfate has been dissolved (Fig. 3-3). When the two copper slabs are connected to a direct electrical current, the current flows through the solution, carrying copper atoms from the slab of blister copper through the liquid to gather on the opposite electrode. The impurities fall out as a sludge at the bottom of the tank. Because gold and silver are often present in copper ore, enough gold and silver can often be recovered from the sludge to pay for the electricity used in the refining. The copper produced by electrolysis is about 99.98 percent pure and can be used where very pure copper is needed.

Fig. 3-3. Electrodes, on which copper will be electrolytically deposited, being lowered into tank at copper refinery. (*Copper Development Association*)

Fig. 3-3a. Schematic view of an electrolytic cell for the final step in refining copper.

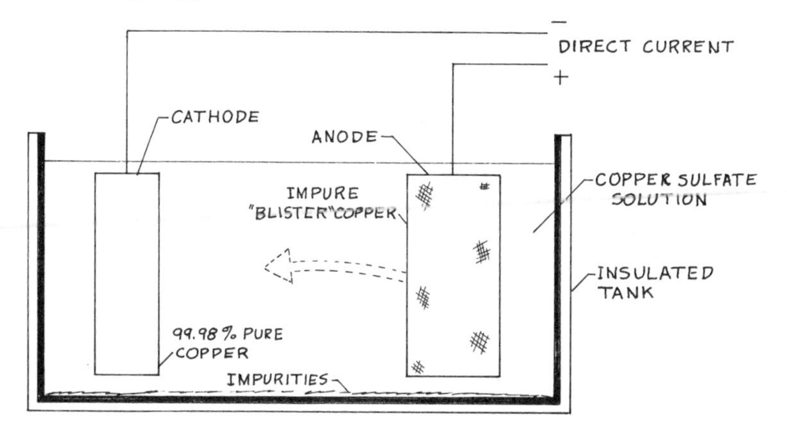

Copper is often mixed with other metals to make alloys such as bronze or brass.

Smelting Aluminum

Iron and copper were first smelted so long ago that we will never know who discovered them or how the discoveries occurred. Aluminum, however, was first isolated as an element in modern times—only about 150 years ago. The first successful commercial process for smelting aluminum was discovered even more recently—less than 100 years ago.

Aluminum is the most plentiful metal in the earth's crust. It is in the sand of the beaches, in nearly all common rocks, and in the clay and dirt under our feet. Before our ancestors began to use metal of any kind, they were making potttery from clay that consisted largely of compounds of aluminum.

In spite of its abundance, aluminum is so difficult to separate from its ore that it was not isolated until 1825. As we have seen in an earlier chapter, aluminum is a very active metal and is therefore never found in nature as pure metal. It is always in compounds, so tightly locked up that neither heat nor chemicals alone can separate out the metal.

Two men are credited with early work on aluminum: the great Danish scientist Hans Christian Oersted, (1777–1851), who first isolated aluminum, and the great German chemist Friedrich Wöhler, who first determined its physical characteristics. Both recognized aluminum as a beautiful and useful metal. But it was so expensive that it could be looked upon only as a curiosity. Around 1860, a pound of aluminum cost as much as $60. A chemical process was later developed to produce aluminum on a larger scale, but it was still too costly to be widely used.

Charles Martin Hall (1863–1914), a young chemistry student in Ohio, became interested in the metal. One of his chemistry books pointed out that every clay bank was a mine of aluminum, but it was so difficult to extract that it cost almost as much as silver. The boy began to think and dream about becoming an inventor and finding an inexpensive way to smelt aluminum.

A few years later, as a chemistry student in Oberlin College, Hall heard his chemistry professor lecture on aluminum. Anyone who invented a process to produce aluminum cheaply, he said, would not only be a great benefactor, but would also make himself a great fortune. Hall began to experiment in earnest in the college laboratories. First he tried the method by which iron and some other metals are smelted—by heating the ore with carbon. This method had been tried earlier by others, but Hall thought that they had not used a high enough temperature. His experiment failed. (Aluminum was so active it would not let the carbon release it from the oxygen that held it in the ore; instead, the aluminum tended to capture the carbon itself to become aluminum carbide.) Next he tried to improve on the expensive chemical method that was then in use—and failed again. He tried high temperatures again, searching for some catalytic agent that would make the carbon-reduction method work. Again, no success.

Having run out of ideas, but still determined to be an inventor, he turned to other metallurgical problems that interested him, not related to aluminum. He experimented, without much success, with a new idea for a battery. He knew that by passing an electric current through water,

water could be broken down, or *electrolyzed,* into its constituents—hydrogen and oxygen. His idea was to reverse the process, introducing a gaseous fuel (such as hydrogen) into one electrode and oxygen into the other, in a way that would allow them to combine to produce water and an electric current. His experiments on this project were never successful, although the idea was successfully developed by others many years later. Electric power was provided in the Apollo spacecraft by "fuel cells" that operated on this principle. During his senior year at college, Hall did some work on the problem of finding a better filament for the incandescent lamp that had been invented a few years earlier by Thomas Edison. When this work yielded no important results, he returned to the aluminum problem.

Perhaps because of his work on the fuel cell, he began to think of using electricity as an agent to produce aluminum by electrolysis; that is, to separate aluminum from a solution of its ore the way hydrogen and oxygen can be electrolyzed from water. He tried to electrolyze water solutions containing aluminum compounds, but produced no aluminum. The project was at that stage when Hall graduated from Oberlin College in 1885.

Back at his family's home, he set up a new laboratory in a large woodshed and set about building batteries and equipment for new experiments. He decided to abandon water solutions of aluminum compounds because he believed that the aluminum, if indeed he produced any in his last experiments, may have reacted immediately with the water and had gone back into a compound. He began to search for a mineral compound that might serve as a solvent for aluminum ore (alumina). The first compound he tried stubbornly stayed a solid no matter how hot he heated it in his furnace. The next few compounds did melt in his furnace, but showed no indication of being able to dissolve alumina.

Then Hall switched to a different compound, sodium aluminum floride, which is a natural mineral known as *cryolite.* The cryolite melted easily into a red-hot liquid. The first step was successful. When Hall stirred pinches of alumina into the melt, the alumina dissolved readily. The second hurdle had been cleared! Now only the third and last step remained—to hook up his batteries to see if electricity would pry the metal aluminum from the ore in the molten bath.

After a time, Hall switched off the current and poured out the cryolite to harden and cool. He broke up the white, rocklike lumps carefully with a hammer. No trace of aluminum could be found, Hall noted, except for a hazy coating on the end of one electrode, which he thought might be aluminum. He repeated the experiment. Still no lumps of aluminum appeared.

Hall decided that the clay crucible he was using might be the wrong material and might be affecting or preventing the reaction he sought. He made a new crucible of carbon and set up his furnace and his batteries once again. It was Tuesday, February 23, 1886. He left the current flowing through the bath for a long time. When the cryolite had cooled and frozen, Hall broke it up with a hammer. Among the pieces of broken rock he found a small silvery button! A few more whacks turned up another bright button. They were metal—light, soft, and malleable. They were aluminum. He had found the secret of unlocking aluminum from the rocks that held it.

The next day Hall made another long run with his equipment and found a small handful of globules of aluminum among the cryolite. Some of the original globules of aluminum produced by Hall's new process have been preserved and are on display today at the Aluminum Company of America (Fig. 3-4).

Hall and a group of associates started a company to produce aluminum. It was called the Pittsburgh Reduction Company. It eventually became the Aluminum Company of America. So useful was aluminum that annual production grew from 10,000 pounds during the company's first year to about 15 million pounds less than 20 years later. In another dozen years, production grew to 44 million pounds. Today aluminum production in the United States alone is more than 8 billion pounds per year.

***Fig. 3-4.* Some of the original globules of aluminum produced by Charles Martin Hall's electrolytic reduction process in 1886. (*ALCOA*)**

Fig. 3-5. Workman breaks the crust on the molten cryolite in a reduction pot in an aluminum smelting plant. (*Reynolds Metals Company*)

Fig. 3-5a. Schematic view of an aluminum-smelting pot. Aluminum is forced out of the electrolyte of alumina and molten cryolite by an electric current and gathers at the bottom of the pot as molten metal.

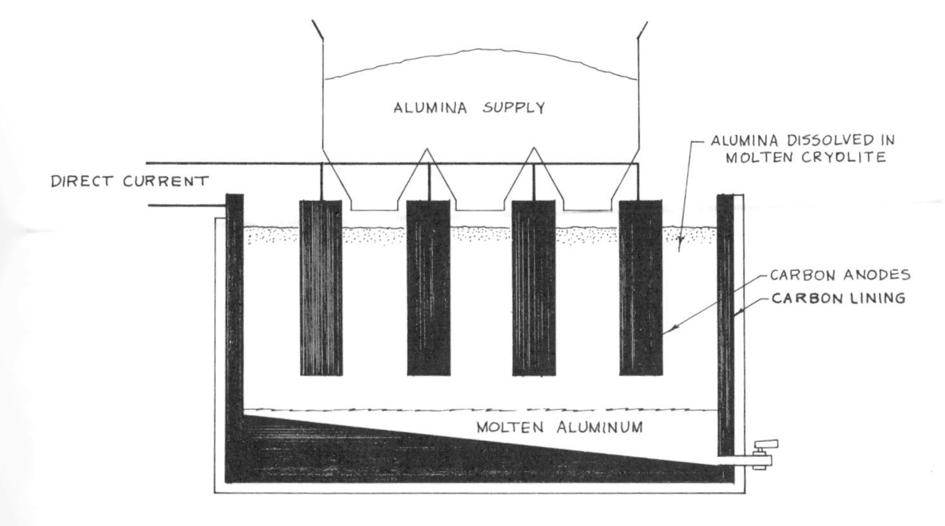

The process of extracting aluminum has since been improved, but it is today essentially the same process developed by Hall. The smelting is done in a string of deep tanks, or "pots," lined with carbon and filled with molten cryolite (Fig. 3-5). The alumina is stirred into this electrolyte. When an electric current is passed through the tank, molten aluminum metal is slowly deposited on the bottom of the cell. The oxygen which had held the metal tied up in the aluminum oxide combines with the carbon from the tank lining and escapes to the atmosphere as carbon dioxide. The pure molten aluminum is drawn off for processing into aluminum products.

Hall, as his chemistry professor had predicted, amassed a great fortune from his discovery. When he died in 1914, he left millions of dollars to his alma mater, Oberlin College, and to other educational organizations throughout the world.

Each of the metals described above is quite different from the others. Iron is strong, but rusts away rapidly. Aluminum is soft and weak. Copper is soft but tough and conducts electricity very well. What makes metals different from each other and different from nonmetals? What *is* metal, anyway? For thousands of years no one had very good answers to these questions. And we don't know all the answers yet. But in the past hundred years, science has found ways to look inside metals to study the atomic and crystal structures. Some of the things that have been learned are the subject of the next two chapters.

CHAPTER 4

INSIDE THE ATOMS OF METALS

Early metalworkers knew how to smelt metal and how to work it. They knew that some metals were stronger than others, and they knew some of the ways to make metal softer or harder. Most of these things they had learned by trial and error after many years of experience. But they really didn't know why their methods worked, or why metal is different from other materials.

During the past century, much has been learned about the chemistry and the atomic structure of all matter. By applying this new knowledge to their work, metallurgists have been able to look inside atoms of metals and learn how they differ from atoms of other materials. They have also learned how atoms are held together in metals, and why metals behave as they do.

All matter is made up of atoms. Although the atoms of each element are different, all atoms are themselves made up of the same basic particles. Each atom of any material is made up of a center part, or nucleus, containing particles called *protons* and *neutrons.* Smaller particles, called electrons, move in orbits in the space around the nucleus. Each proton in the nucleus has a positive electrical charge, while each electron orbiting the nucleus has an equal and opposite (negative) charge. Particles with negative and positive charges are attracted to each other by an electrical force. Each proton in the nucleus attracts and holds one electron in orbit.

The number of protons in the nucleus makes an atom the particular element that it is. For example, the simplest atom in the Universe consists of one proton and one electron (Fig. 4-1). It is a gas called hydrogen. The next heavier element has two protons, two electrons, and usually two neutrons. This element is also a lightweight gas, four times as heavy as hydrogen. It is called helium. The other 90 natural elements (that is, elements found in nature) are made up of various numbers of protons, electrons, and neutrons. The heaviest natural atom has 92 protons, 92 electrons, and weighs about 236 times as much as a hydrogen atom. It is the metal uranium. (Uranium also has 146 neutrons, but these are unimportant for our discussion.)

***Fig. 4-1.* The number of protons in the nucleus makes an atom the particular element that it is. Each proton holds one electron in orbit around the nucleus.**

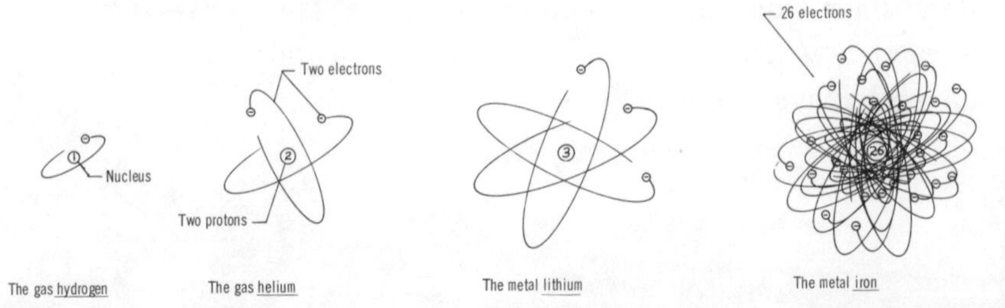

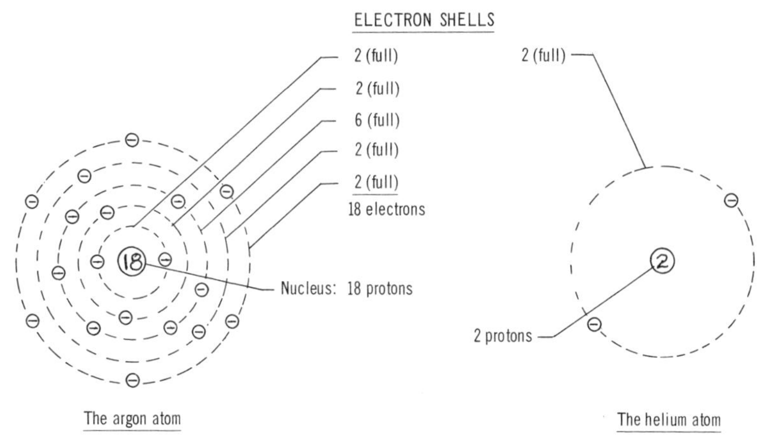

***Fig. 4-2.* Both argon and helium are "inert" gases because they have their outer electron shells filled to capacity. They neither need electrons nor have any to lend.**

Electrons in orbit around the nucleus are arranged in layers or "shells." (Niels Bohr, a Danish physicist, first introduced the idea that orbiting electrons were arranged in "shells." Later, it was found that these shells are arranged in groups, have shapes other than circular, etc. For the purposes of this book, the simplest schematic shell arrangements, depicted as individual circular orbits, will be used.) Each shell is located at a certain distance from the nucleus and can contain only a certain maximum number of electrons. It may contain fewer but not more electrons than this number.

The number of electrons in each shell, compared with the number required for a "full" shell, determines how the element joins with other elements; that is, how it behaves *chemically*. When an atom has fewer electrons than are required for a full outer shell, it is more likely to combine with another atom—one that will provide or share electrons to fill the vacancy in the outer shell.

Figure 4-2 shows the electron shell arrangements for the element argon, a gas found in very small quantities in ordinary air. All of argon's electron shells are full; that is, they contain the maximum number that can occupy each shell. Because the argon atom has no electron shells that are only partly full, it has neither any electrons to give away nor any need for more. Consequently, argon will not easily combine chemically with any other element. For this reason, we call it an *inert element* (gas). Helium (also shown on Fig. 4-2) is also an inert element (gas) because its only electron shell is full.

Chemically, the gas hydrogen behaves quite differently from an inert gas because its electron shell is lacking one electron (Fig. 4-3). Hydrogen is a very "active" element; that is, it combines readily with many elements to make chemical compounds. A hydrogen atom will ordinarily not exist by itself, but will join with another hydrogen atom in order that they may share their electrons (center sketch of Fig. 4-3). As separate atoms, each is short one electron, but by joining together they can share their electrons and have two whirling around both nuclei. Together two hydrogen atoms make up a hydrogen *molecule.* (A molecule is the smallest particle of any substance which can exist and still have all the chemical properties of that substance.)

One of the elements with which hydrogen will combine

Fig. 4-3.* Atoms that do not have full outer electron shells may share electrons with atoms of other elements and form a *covalent bond.

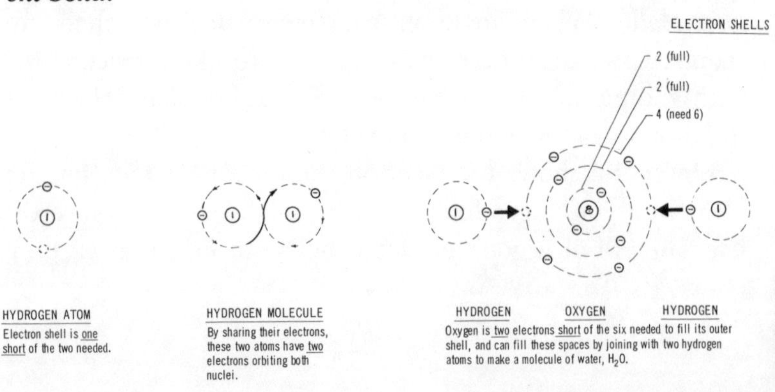

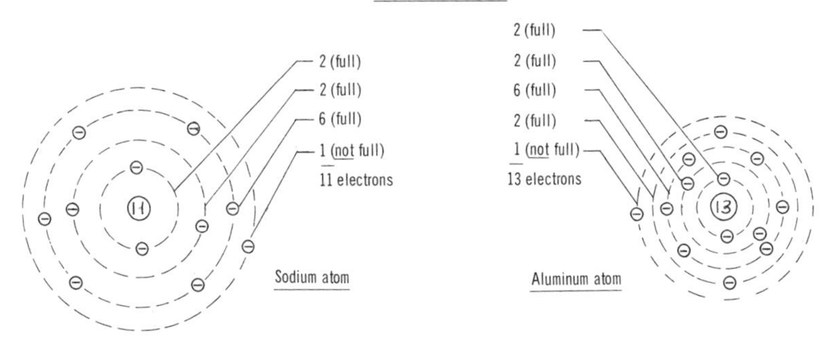

***Fig. 4-4.* Metal atoms tend to have outer electron shells that are only partially filled. These "loose" outer electrons are available for sharing.**

easily is oxygen. The element oxygen also behaves differently from an inert gas because it has only four electrons in its outer shell—which is two short of filling that shell. Oxygen, like hydrogen, is a chemically active element in need of electrons. When the oxygen atom combines with two atoms of hydrogen, as shown in the sketch, it acquires the two electrons needed to fill these empty places, and a molecule of water (H_2O) is formed. The electron bond which holds two atoms of hydrogen together in the hydrogen molecule, and which holds hydrogen and oxygen together in the water molecule is due to the sharing of electrons and is called a *covalent bond.* (The bond that holds a metal in the compounds from which it is smelted is usually a covalent bond.)

If we look at the atoms of metals, we will find that they tend to have partially-filled electron shells. You can compare the electron structures of sodium and aluminum, shown in Fig. 4-4, with the structure of argon in Fig. 4-2. Sodium has only one electron in its outer shell where two are required for a full shell. Aluminum has only one electron in its outer shell where six are required for that shell to be full. In each case, the single electron is more loosely held within the atom than are electrons in full shells and so it is more free to be loaned or shared with other atoms.

When a metal combines chemically with another element to form a compound, the metal may *lend* an electron to the other element rather than sharing it. For example, when the metal sodium joins with chlorine to make the compound sodium chloride (NaCl), or common table salt, the sodium lends its loose electron to chlorine, which is short one electron in its outer shell (Fig. 4-5). The loss of the electron leaves the sodium atom with a *positive* charge, while the gain of an electron gives the chlorine atom a *negative* charge.

These charged atoms are called *ions*. The positive ion (the sodium) is attracted to a negative ion (the chlorine) by an electrical force. This force can be demonstrated with a plastic comb and small bits of paper. Running the comb through your hair or rubbing it on a piece of wool cloth will give the comb an electric charge. Bits of paper can now be picked up by the comb. The force that holds the paper to the comb is similar to that felt between positively and negatively charged ions. In chemistry, this kind of force forms the bonding between atoms (ions) which is called an *ionic bond*.

When sodium chloride (NaCl) forms a solid crystal of table salt, the sodium ions (positively charged atoms) and

***Fig. 4-5.* The chlorine atom, which is missing one electron in its outer shell, captures a loose electron from the sodium atom. The attraction between the chlorine ion (−) and the sodium ion (+) is the *ionic bond* that holds the NaCl molecule together.**

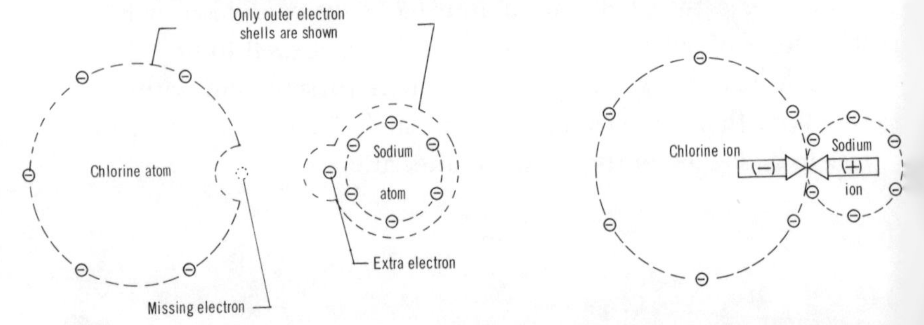

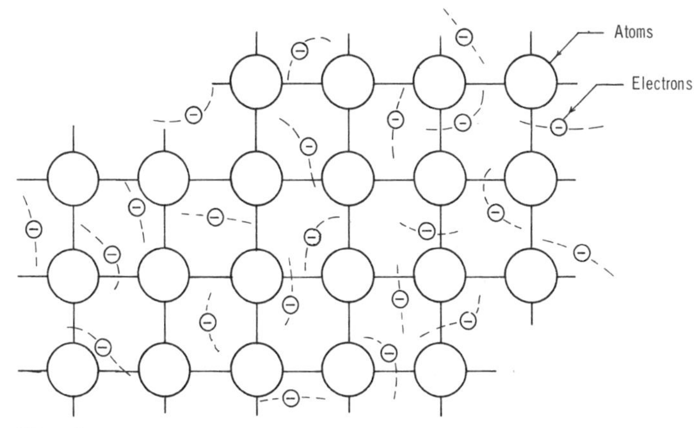

Fig. 4-6.* Atoms of metal in a crystal lattice are held together by electrons circulating freely among the atoms, forming a *metallic bond.

chlorine ions (negatively charged atoms) are held together in the crystal by this electrical force. The atoms must be very close to each other in order for this force to be effective. If an outside force pulling on the crystal is strong enough to move these ions far enough away from each other, this attractive force is reduced to where the "bond" will fail. Because they have these kinds of bonds, crystals of some metal compounds, like sodium chloride, and crystals of other nonmetallic compounds are brittle and will break apart easily and suddenly.

When atoms of a metal join *each other* to form a crystal of solid metal, the loose electrons behave differently. They produce a third kind of bond called a *metallic bond* (Fig. 4-6). The loosely held electrons do not stay attached to any particular atoms, but circulate freely through the crystal. The attraction between the negative charge of the free electrons and the positive charge of the metal atoms is the "glue" that holds a metal crystal together. This glue gives the metal crystal its ability to be squeezed or bent out of shape without breaking. The atoms of metal can move a greater distance from each other without breaking their bonds because the circulating electrons can move around wherever they are needed to hold the atoms together.

To look at the electron bonds in another way, we can visualize the ionic bond holding atoms together in a crystal of a compound (such as salt) to be like a slender thread of glass. The thread is strong, but if the atoms are pulled only a small distance apart, the brittle thread snaps, and the crystal breaks. On the other hand, the metallic bond holding atoms together in a crystal of metal is like a strong rubber band. The atoms can be pulled much farther apart without breaking, because the rubber band is flexible and can stretch as needed to hold the atoms together.

The most important feature of metals, then, is not only that they are so strong that they will not break easily, but also that they can be pulled or squeezed or twisted into a very different shape *without* breaking. This characteristic is called *ductility*. A metal (such as aluminum, for example) can be rolled from a heavy slab into foil as thin as paper, all because the flexible electron-glue holds the atoms together, even though they move and change positions over great distances.

The electrons that circulate through the crystals also give metal another of its important characteristics—the ability to conduct electricity. An electric current is a flow of electrons along a conductor. When an electrical voltage is applied to a metal conductor, it sends the loosely held electrons scurrying along through the crystals to carry the current. In nonmetallic materials, such as glass, there are no loose electrons available because these materials are stable compounds with filled electron shells. Material that will not carry an electric current is called an *insulator*. The electrons are so tightly locked up in the atoms of an insulator that a very large electrical voltage would be needed to break them loose. With some insulators, the electrical voltage needed to break electrons free is so large that the material itself will shatter when the electrons are finally forced to move.

The loosely held electrons also give metals good heat-conducting qualities. The atoms of any solid substance move rapidly back and forth—or "vibrate"—constantly. When one end of a bar of metal is heated, the atoms at the hot end vibrate more rapidly. The circulating electrons help to carry the vibrational energy quickly from the hot end to cooler parts of the bar. In a nonmetal, which has no free electrons, heat energy can be transmitted along the bar only by the vibrating atoms banging into each other; consequently, heat is conducted much more slowly in these materials.

By looking inside the metal atom, we have learned that it is the loosely held electrons that make metals different from nonmetals. In order to learn what happens inside a metal when it is stretched or bent, and why one metal is stronger or harder than another, we must look at how metal crystals are formed and how they behave.

CHAPTER 5

INSIDE METAL CRYSTALS

In the early days of the science of metallurgy, no one was sure just how the atoms were arranged in solid metal. There were two main possibilities. The atoms could be arranged in rigid crystal structures, as they are in salt, ice, and the rocks from which metals are smelted. Or, the atoms could be simply piled on top of each other, in an unordered arrangement, like those of a liquid. Many solids, such as wax, pitch, and glass, have this *amorphous* form. They are really supercooled liquids; that is, gooey liquids that become more and more pasty as they cool until they are finally so stiff that they appear to be a solid. Although the atoms in such a solid are not free enough to permit the material to pour, they are not locked into any kind of permanent structure. A piece of amorphous material such as wax, taffy candy, or pitch left sticking out over the edge of a table will eventually bend and droop over the edge, especially if it is warmed a bit by the sun. Even glass, which is our most common amorphous material, will bend and droop if it feels a force for a long time —although it might take a very long time (many years) for the deflection to be noticeable.

There was evidence for and against both the rigid-crystal and the unordered-arrangement ideas, but for a time the most common belief was that metal is an amorphous solid. Most crystalline materials, after all, were brittle, weak, and crumbly. Metal appeared to be quite the opposite. One metallurgist noted that solid lead metal could be squeezed out through a hole, if pushed with enough force, like toothpaste from a tube. Solid metal, he reasoned, flows as though it is an amorphous material like wax or pitch.

The idea that metals were amorphous solids held for a time, but the evidence against it began to grow. Pitch and taffy seemed ductile and would bend easily, like metal, if bent slowly. But they would snap if bent too quickly and would shatter into pieces if struck by a hammer. Most metals didn't act that way at all. Also, the broken edge of a piece of pitch, taffy, or glass has a shiny, glossy look. If you bend a metal nail back and forth until it breaks, and examine the broken end, or look closely at a piece of broken cast iron, you will see that broken metal looks quite different. A magnifying glass will help, but even with the naked eye you may be able to see that the break has a rough, grainy appearance. These grains, which look like very fine sand or sugar, give strong evidence that metal is made up of crystals.

Metallurgists slowly came to believe that although metals act differently from most common crystalline materials, they

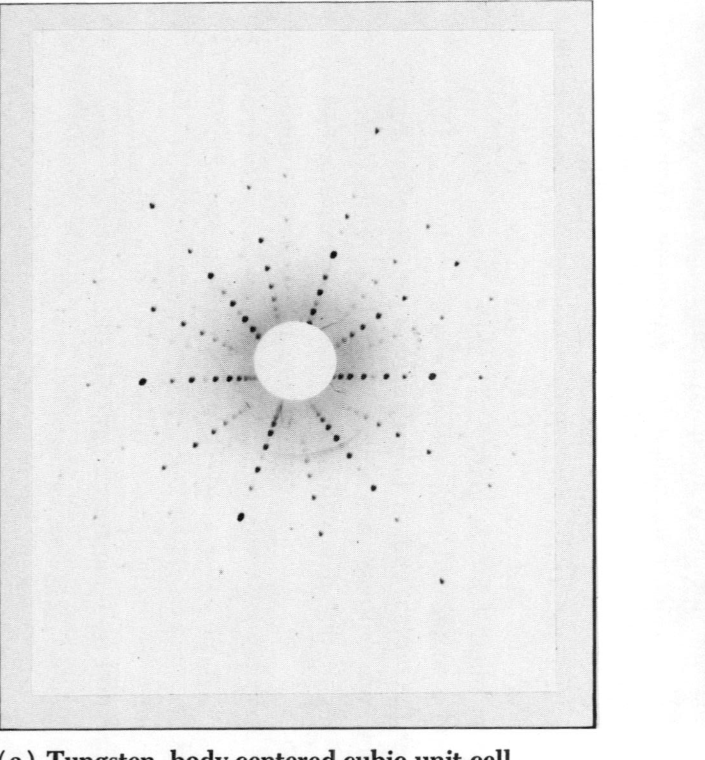

(a) Tungsten, body-centered-cubic unit cell.

(b) Germanium, diamond unit cell.

Fig. 5-1. X-ray patterns of crystals of two metals. (*C. Becktoldt and M. Kuriyama, National Bureau of Standards*)

nevertheless are made up of crystals. The argument raged for many years and was finally decided when metals were studied by means of X-rays.

X-rays are most familiar to us as a method of seeing inside the human body. The rays pass right through the body to make a shadowy photograph of the organs encountered on the way. Bones, organs, and muscle tissue show up as various shades of light and dark because the X-rays pass through some of them more easily than others. We wouldn't learn much from X-rays of metal made in this way, because a metal crystal is much too small to show up in such a photograph. Instead, the X-ray beam is directed at a thin sample of metal. The rays pass through, or are reflected, or are bent, depending upon whether they collide with the atoms of metal, and are recorded on film as a complicated pattern of spots and lines (Fig. 5-1). From measurements of this pattern, calculations can be made to find the spacing and arrangement of atoms in the metal.

X-rays gave final proof that atoms of solid metal are not just lying together in an amorphous heap, but are arranged in a very precise pattern, or "crystal lattice" (Fig. 5-2). We now know what the crystal lattice looks like for most metals.

The smallest portion of a crystal is called a "unit cell." Fourteen different kinds of unit cells have been found. Three of the commonest crystal structures for metal are shown in Fig. 5-3. The two at left are unit cells, while the structure at right is a combination of three simpler unit cells. Some of the metals whose crystals normally have these shapes are shown on the figure.

Metal atoms in a body-centered cubic cell are simply arranged in a cube—like an imaginary square box with an atom located at each corner and one in the exact center

Fig. 5-2. **Ping-pong ball model of the arrangement of atoms in a close-packed-hexagonal lattice. (*Reprinted by permission from* The Nature of Metals *by Bruce A. Rogers, published by Iowa State University Press.*)**

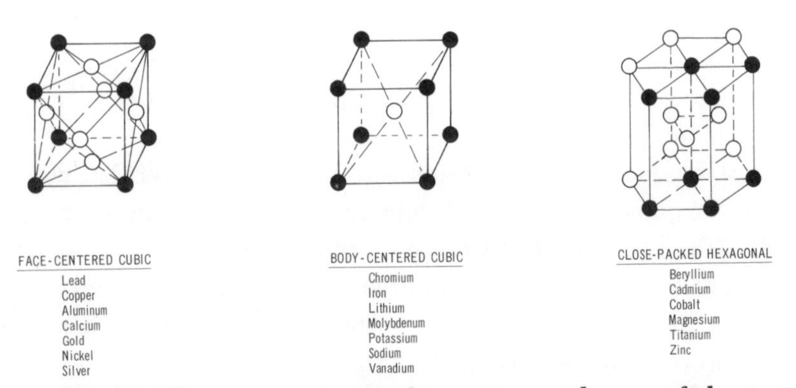

Fig. 5-3. **Three common crystal structures and some of the metals that have these forms of lattice. For clarity, the atoms that form the basic unit cell shape are shown in black.**

of the box. The face-centered cubic cell has an atom at each corner and one at the center of each face of the imaginary box. The hexagonal structure is made up of a simpler unit cell and has an atom at each corner and three more located inside the lattice as shown in the illustration.

Each atom attaches itself to its neighbors as the molten metal solidifies—or freezes—into unit cells of its own kind. The atoms have been drawn with a great deal of space between them in order that their position and the shape of the unit cell can be more clearly shown. If you could see individual atoms, they might appear more like a pile of oranges on a fruit stand, nestled against each other in orderly rows, layer on top of layer, to form the regular crystal lattice.

Some metals have different crystal structures at different temperatures. The unit cell of sodium is normally a body-centered cubic structure, but at low temperatures this unit cell can change to the hexagonal close-packed structure. Iron can have two different cubic arrangements. At ordinary temperatures the unit cell is a body-centered cubic structure, but at high temperatures the iron atoms can rearrange themselves to make the unit cell a face-centered cubic structure.

The unit cell in any metal is, like the 8 or more atoms that make it up, extremely small. One gram of copper may contain more than 2,000,000,000,000,000,000,000 (two thousand billion billion) unit cells. The kind of unit cell that a metal forms when it crystallizes determines how closely packed its atoms are and is therefore one factor determining the specific gravity (weight per unit volume) of the metal. The face-centered cubic and hexagonal close-packed structures give the densest packing of atoms (atoms are usually pictured as being spherical in shape). The other facts that determine the specific gravity of a substance are the weight of its atoms (atomic weight) and the size of its atom (atomic radius). We can see the effects of these properties if we compare the metals uranium, gold, and osmium in Fig. 5-4.

Uranium has the heaviest atom of the three (and the heaviest atom of *any* natural element). It has a large atomic radius and a simple cubic crystal structure which is a relatively loosely packed arrangement. Consequently, uranium has a lower specific gravity than either gold or osmium.

Both osmium and gold have tightly-packed crystal structures. Although gold has a heavier atom, the smaller atomic radius of osmium gives osmium a smaller unit cell and more atoms in a given volume of metal. As a result, osmium has a specific gravity greater than that of gold. Osmium's specific gravity is, in fact, the greatest of any natural element, 22.6 times the weight of water. The element with the

***Fig. 5-4.* Crystal structure and specific gravity for four metals including *osmium*, the densest known element, and *lithium*, the lightest solid element.**

Metal	*Crystal Structure*	*Approximate Atomic Weight*	*Atomic Radius*	*Specific Gravity (Water = 1)*
Uranium	Cubic	238	1.5	19.07
Gold	Face-centered Cubic	197	1.44	19.3
Osmium	Hexagonal Close-packed	190	1.35	22.6
Lithium	Body-centered Cubic	7	1.52	.53

smallest specific gravity of any solid is the metal lithium. With a specific gravity of 0.53, lithium weighs a little more than half as much as water.

You can see how crystals grow by looking at ice forming on a puddle or on a window pane. The water does not freeze all at once (Fig. 5-5). Instead, crystals appear at a few points on the surface, then grow and spread, branching first this way, then that. Figure 5-6 shows schematically that the growing crystals of the lattices usually do not match, because each crystal has its lattice lined up in a different direction. Consequently, where two crystals meet, the atoms at the crystal boundaries join each other in irregular ways. The crystal lattice at these boundaries is bent and twisted out of its usual shape, which causes the grain boundaries to be the strongest and hardest part of the metal.

Because metal solidifies in this way, a piece of solid metal is never one large crystal but is always divided into many small crystals of various sizes and shapes. A chunk of metal

***Fig. 5-5.* These ice crystals on a window pane illustrate how metal crystals form and grow as metal solidifies. (*Norman F. Smith*)**

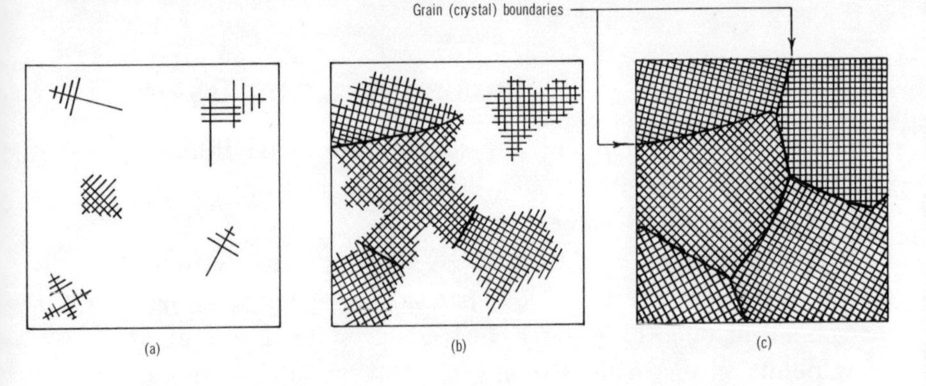

Fig. 5-6. **When metal solidifies, crystals begin to form with lattices lined up in different directions (a). The crystals meet (b) and continue to grow until the metal is solid (c).**

is not like an orderly pile of oranges that goes on and on, row after row, as one crystal would do, but is more like a group of orderly piles (or crystals), one running this way, another that way, with oranges fitting in odd ways between the groups.

Although a grainy texture may be visible to the naked eye in a broken piece of metal, the crystals in metal are so small that a microscope is needed to see them clearly. If a piece of metal were polished and dipped in an acid that would eat away some of the metal between the crystals, you would see a definite structure of crystals, or grains, through a microscope (Fig. 5-7). A grain (or crystal) of hard metal may be from 0.01 to 0.1mm in diameter, while a grain of soft metal may be 10mm in diameter or even larger.

Individual crystals of zinc are usually visible on a new galvanized bucket. When the bucket was dipped into the molten zinc metal, a thin layer of zinc froze to the surface in crystals so large that you can easily see their shapes and patterns with the naked eye. The size and shape of metal crystals have a great deal to do with the hardness and strength of the metal, as we shall see in a later chapter.

Fig. 5-7. **The individual grains of metal can be seen in this microscopic photograph of annealed cartridge brass. Surface has been polished and etched by acid. Magnified 75 times.**

Although crystals are orderly structures, not all atoms will be lined up and bonded in perfect order. Each crystal contains large numbers of irregularities or defects. One kind of defect, called a dislocation, is caused by an incomplete row of atoms (Fig. 5-8). Another kind of very simple defect is caused by an atom missing here and there in the lattice. This kind of defect is known as a "vacancy."

What happens inside these metal crystals when the metal is stretched, bent, or hammered? If the force applied to the metal is not too great, the movement of the atoms will be small, and they will move back to their original positions when the force is removed. Remember the oranges piled in layers on a fruit stand. If we push gently against a layer of oranges, it may move slightly. When we stop pushing, the

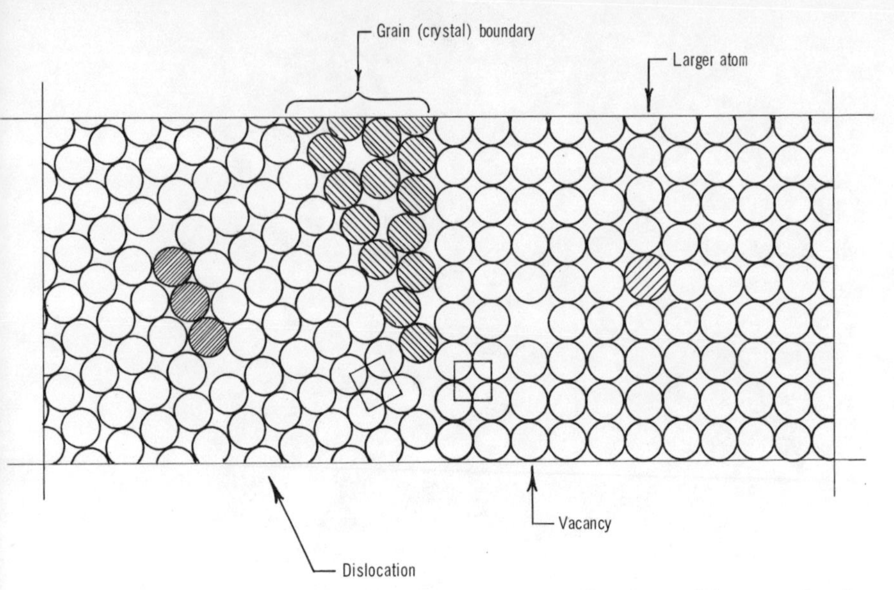

***Fig. 5-8.* Some irregularities in atomic lattices. (To see the irregularities more clearly, view from the bottom of the page in the direction of the arrows.)**

oranges will move back to their original positions. Most of the things we use are designed so that the metal deflects in this way. This process is called *elastic deformation.* A metal screen-door spring is designed to stretch elastically in use, always returning to its original shape.

If a force applied to metal is great enough to move the atoms out of position, they will "slip" permanently into a different position. This process is called *plastic deformation* (Fig. 5-9). If we push harder on one layer of those fruit-stand oranges, it will slide over the layer below and remain in this new position when we stop pushing. This is what happens to the metal atoms when a piece of metal is bent and does not return to its original shape. A screen-door spring that is stretched a little too far will return only part way to its original shape when it is released. The atoms in some of its crystals have slipped to new positions and this has left the spring permanently deformed.

When slips occur, not every layer of the crystal slips—only a layer here and there. But each slip is usually a big one, carrying an atom past dozens or even hundreds of other atoms before it stops. Figure 5-10 shows actual slip planes in crystals of cold-worked nickel-aluminum alloy, magnified 1000 times. Slipping occurs mainly in the areas where there are weaknesses due to vacancies and dislocations in the crystal. We can imagine that a layer of oranges that has one or more oranges missing in a row will move a little more easily. As it moves, the vacancy may be filled in, or it may just be bumped to another place in the crystal. Slip occurs easily in certain crystal planes but not in other crystal planes that are aligned in different directions.

Another analogy to crystal slip at dislocations is found in moving a small rug on a floor. If the rug is perfectly flat, it may be difficult to slide. But if it has wrinkles and waves in its surface (which correspond to dislocations), the rug between these areas will slide easily.

Because dislocations are the sites for crystal slip, we might expect that perfect metal crystals that have no dislocations might be a great deal stronger than ordinary metal. Research on single crystals or "whiskers" of metal have shown this

***Fig. 5-9.* A force (arrow) deforms metal by causing its crystals to slip. Slip deforms the crystal boundaries, which makes the metal harder and stronger.**

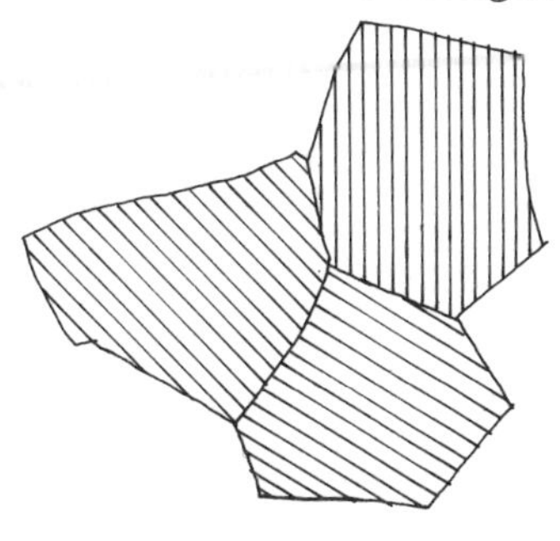

***Fig. 5-10.* Slip planes are clearly visible in these crystals of cold-worked nickel-aluminum alloy which are shown magnified 700 times. In actual size, the slip-planes are about 1/40,000-inch apart. (*W. Willard, National Bureau of Standards*)**

to be true. The strength of such samples has been measured to be from a few dozen times to several thousand times as strong as ordinary metal. No one has yet found out how to produce such crystals in sizes large enough to be useful. If this can be done, a great deal less metal could be used to carry the same load. For the time being, however, we must deal with real metals, which contain thousands and millions of dislocations that limit the strength of the metal.

One of the surprising things about slip is that a metal is not weaker after it slips, but *stronger* than before. Slipping pulls crystals out of shape and pushes them up against other crystals; as a result it will be more difficult for them to slip again. Because a larger force is needed to make them slip next time, we can say that metal is stronger after it has been deformed. If we were to continue to stretch or bend a piece of metal, each slip would make the metal stronger and harder. Finally the metal would become so hard and brittle that it will break rather than deform further. If we keep stretching the screen-door spring far enough, we will have only a long, crooked, stiff wire that will have very little spring to it and will eventually snap.

In an earlier chapter, it was pointed out that metals conduct electricity because they have free electrons that can move about readily through the crystals. Because defects in the crystal, impurities, and strains due to cold-working hamper the movement of electrons, they can all decrease conductivity of a metal.

Having made the point that atoms of metal are locked rather firmly into a crystal lattice, we must now modify that view somewhat. Even when locked into a lattice, atoms are not really standing perfectly still, but are vibrating constantly. If the temperature of the metal is increased, the atoms will vibrate more vigorously, until they finally break loose and move freely as individuals in a molten mass. This "breaking loose" point is the *melting temperature*. As the temperature is increased still further, the motion with which the atoms jostle each other in the liquid increases steadily until it becomes so great that the atoms begin to break loose from the liquid and fly off into space—as a gas. This is the *vaporization temperature*.

If, on the other hand, the temperature of a solid metal were continuously decreased, the vibration of the atoms in the crystal would decrease until it finally came to a complete stop. The temperature at which atomic vibration would stop is the point at which all heat has been removed from the metal. This temperature is called *absolute zero*. No lower temperature is possible. Absolute zero is –460°F (–273°C).

(Absolute zero can be closely approached but can never be reached in the laboratory.)

Atoms of all materials are thus in constant motion at all times if their temperature is above absolute zero. At ordinary temperatures atoms of metal not only vibrate in place but can actually jump from one position to another in a crystal lattice. Such movement was shown in an experiment in which a slab of lead was clamped tightly to a slab of gold for four years. When the slabs were again separated, gold atoms were found inside the block of lead as much as 5⁄16 inch below the surface! It appears that atoms can jump into vacancies in crystal lattices and perhaps even change places with each other. Other experiments have shown that the atoms in a crystal of a simple (pure) metal can move around in the same way.

In solid metal at ordinary temperatures, this movement of atoms from one place to another, called *diffusion,* is usually a very slow process, as it was in the lead-gold experiment. But if we raise the temperature of a metal to between 1⁄3 and 1⁄2 of the melting temperature, the atoms can easily move far enough to adjust the lattice of the solid metal. We can use this atomic movement to make a metal soft again after it has become hard and brittle through crystal slip. This process is called *annealing.* When the hard metal is heated, the first atoms to jump to other positions are those that had been pushed out of place at the crystal boundaries. As they move, the crystals relax and get rid of the twists, kinks, and strains that slip has produced. Some of the small crystals may combine by realigning their lattices to form larger crystals (Fig. 5-11). After the metal has cooled, it will be soft again, and crystal slip can occur more easily. Although they did not know why it worked, ancient metalworkers

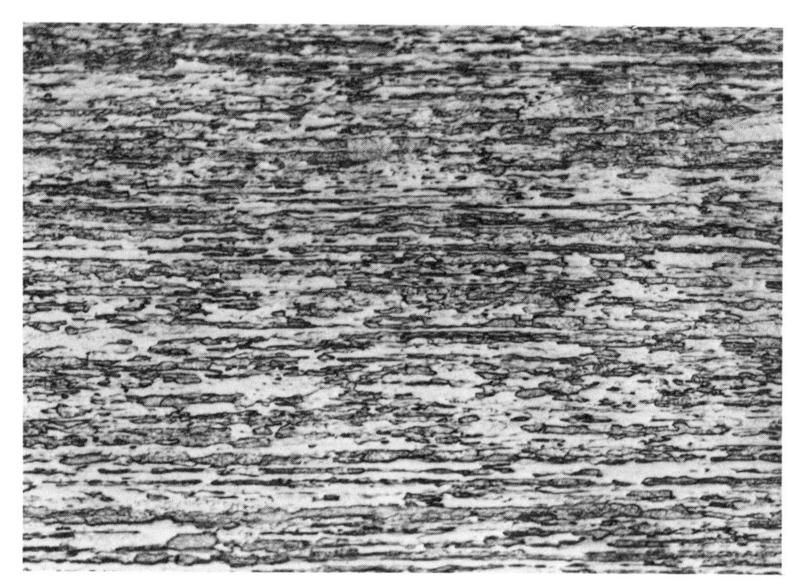

(a) Cold worked molybdenum; with small, flattened grains; metal is hard. Magnified 200 times.

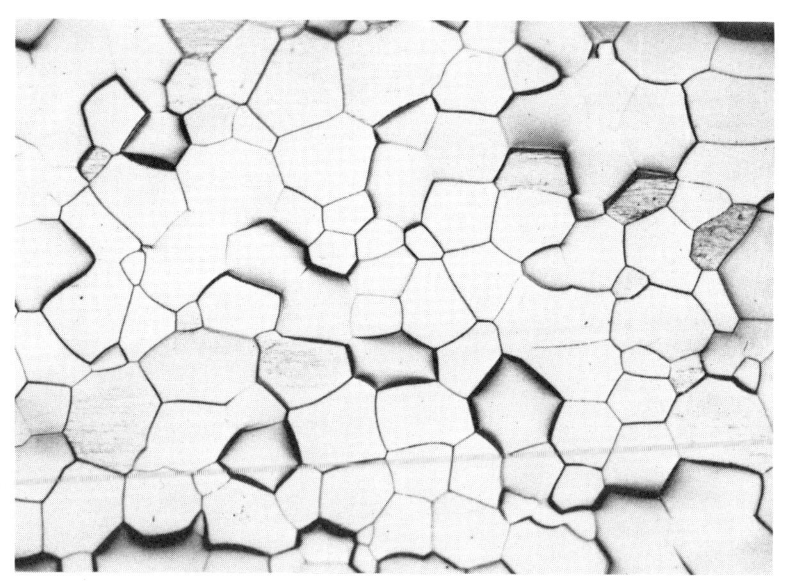

(b) Same metal annealed; with long, relaxed grains; metal is now soft. Magnified 200 times.

Fig. 5-11. Annealing softens metal by changing its grain structure. (*C. Brady, National Bureau of Standards*)

used this method of annealing to soften metal in their forges that had become hard through hammering.

The lead-gold experiment described earlier suggests that two atoms of different sizes, such as lead and gold, can take positions beside each other in the same crystal lattice. What should we expect to happen if we were to mix two or more metals in the molten state? This is an extremely important question, because most metals are not used in their pure form but rather mixed with others in combinations called *alloys*. We can go back to the fruitstand analogy and ask what would happen to the crystal structure if we were to replace an orange here and there with a grapefruit, a lemon, or a cherry. Some answers to this question will be found in the next chapter.

CHAPTER 6

INSIDE METAL ALLOYS

When early metalworkers first smelted bronze, they probably believed they had discovered a wholly new kind of metal. Bronze was a slightly different color and much stronger, harder, and more durable than copper. Actually, they had stumbled upon the very simple but important process called *alloying*. They had mixed with their copper, deliberately or accidently, a small amount of the metal tin. (Other alloying elements may have been used before tin, but tin made the first really successful alloy with copper.) Both copper and tin are so soft and weak that they are not very useful as pure metals, especially where hardness or strength is important. But mixed together, these two metals make the alloy bronze, which is much stronger, harder, and more durable than either metal alone.

Few metals today are used in their pure form for purposes where strength is important. Most are too soft and weak to be of practical use. Three of our most basic metals, iron, copper, and aluminum, are nearly always used in combination with other metals in hundreds of different alloys. Developing new alloys to serve new purposes is one of the tasks of the *physical metallurgist.*

Why do two soft, weak metals make an alloy stronger than either metal alone? The answer lies inside the alloy, in the way the atoms of the alloying metal locate themselves in the crystals of the parent metal, and affect the slipping of these crystals. (*Parent metal* means the metal to which the small amount of *alloying metal* is added.)

To make a metal alloy, we must mix two or more metals together while they are melted. How different metals behave when mixed together depends upon how well they *dissolve* in each other. Some metals are completely *insoluble* in each other. That is, if they are mixed together when molten, they will not stay mixed, but will separate out in layers when they are left standing. A familiar example of two substances that are insoluble in each other is oil and water. Oil and water will not stay mixed, but will settle out in two separate layers as soon as you stop stirring them. The atoms of each substance, we might say, are attracted only to their own kind and not to the other kind of atom. Gravity will cause the lighter atoms to float to the top and the heavier to sink to the bottom.

An example of insoluble metals is found in lead and copper. Molten lead will not dissolve in molten copper, but will remain in a separate layer. If stirred into the melt, the lead may appear to mix with the copper. But if you were to examine a sample of the solidified mixture under a microscope, you would see tiny globs of lead scattered throughout the copper.

Some metals are completely *soluble* in each other. That is, when the metals are mixed together in the molten condition, their atoms stay mixed so completely in each crystal when the metal solidifies that we can detect no globs or crystals of either pure metal. Alcohol and water are familiar materials that are completely soluble in each other. They mix completely, no matter how much of either substance is present. Among the metals, silver and gold are completely soluble in each other, as are copper and nickel.

Most metals, however, lie somewhere between these two extremes of complete insolubility and complete solubility. They will dissolve only limited amounts of other metals—perhaps a large amount of one metal, but only a small amount of another.

Sugar and water are an example of this kind of limited solubility. If you stir one spoonful after another of sugar into water, the first few will dissolve rapidly. If you continue, the mixture will reach a point where no more sugar will dissolve. No matter how much sugar you add, it will remain as grains in the bottom of the cup.

Metals will dissolve in each other more easily when they are liquids, because in liquids the atoms are simply piled on top of one another, helter skelter, and have no firm structure (Fig. 6-1). But when an alloy begins to freeze into solid metal, we may find that the two metals have *limited solubility;* that is, the atoms may not stay mixed when they must begin to arrange themselves into a crystal structure. As we might expect, atoms that are too different from the rest tend to be pushed out as the crystal freezes.

If we use a microscope to examine the crystals of such an alloy, we may find some crystals that are nearly pure metal of one kind or the other, crystals that are rich in one metal or the other, and crystals that are complete mixtures of the

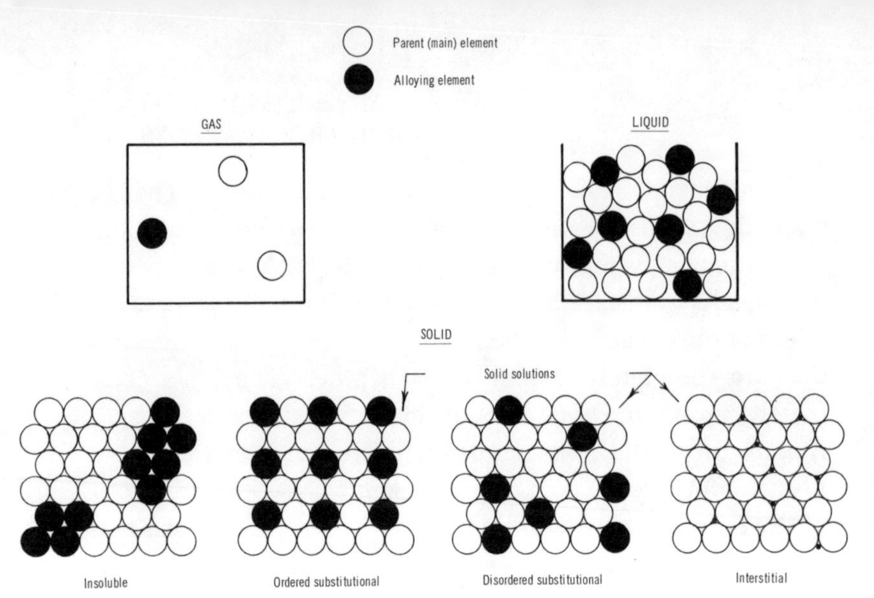

***Fig. 6-1.* The arrangement of atoms in an alloy when it is a gas, a liquid, and a solid.**

two metals. The complete mixture is called a *solid solution.* It consists of crystals in which atoms of both metals exist in the same lattice, side by side.

The commonest kind of solid solution is the *substitutional,* wherein the atoms of the alloying element simply substitute themselves for the regular atoms in the crystal. Two kinds of substitutional solid solutions are possible, as diagrammed in the bottom sketches of Fig. 6-1: the *ordered,* where the alloying atoms take up the same position in each unit cell, and the *disordered,* where atoms of the alloying element are scattered randomly through the crystal lattices. For most metals, these two arrangements seem to have about the same effect on the hardness or strength of the metal.

A less common kind of solid solution is the *interstitial,* where the alloying atoms (because they are small) are scattered in the spaces between the other atoms in the lattice. Carbon, having a small atom, takes up interstitial locations in steel.

Three factors that determine how well one metal will form a solid solution with another are: a) the size of the atoms of the two metals, b) the chemical "attraction" between the two different atoms, and c) whether the two metals have the same crystal structure (unit cell). Atoms of similar size will mix more easily than atoms of different size. Silver and gold (which, as mentioned earlier, are completely soluble in each other) have atoms of almost the same size and the same unit cell. On the other hand, lead has a very large atom; consequently, it does not alloy easily with most other metals—in fact, it is completely insoluble in some, such as copper and iron. Atoms with a chemical attraction for each other may join chemically to form a compound instead of simply mixing as an alloy.

Because each metal has atoms of a different size and different chemical characteristics, the ways in which metals form alloys with each other are complicated. No simple rule has been found that will explain in advance exactly what will happen when two metals are mixed together, or what the alloy will be like. But by looking inside the metal crystals, we can see some of the factors that give an alloy its characteristics.

Solid-solution alloys, that is, alloys in which the atoms mix freely in the crystal lattice, are among the most important alloys. Most brasses and bronzes and some steels are solid-solution alloys. The fruitstand analogy, with oranges neatly piled row-on-row, layer-on-layer, like atoms in a crystal, will again be helpful in describing what happens inside solid-solution alloys. Let's imagine that we mix other atoms of different sizes, represented by grapefruit and lemons, into the pile of oranges. The place of an orange here and there will be occupied by a grapefruit or a lemon. The rows and layers of oranges will now be distorted—humped up where a larger grapefruit sits and sagged down where a smaller lemon is located. If we think about pushing one layer of the pile until it slides over the other, we will see that a harder push would be required because the rows are no longer neat and orderly.

The alloying atoms in a solid-solution alloy act in much the same way as the grapefruit and lemons. Atoms of the alloying metal substitute for, or replace, atoms of the original metal in the crystal lattice (Fig. 6-2). If the added atoms are nearly the same size, the alloy may not be greatly different from the original metal. If the atoms are either a larger or smaller size, they will distort the lattice and make slip more difficult, so that the alloy is stronger and harder than the original metal.

Now suppose that we mix still smaller fruits, like cherries, into the pile of oranges. The cherries would not be large enough to take the place of oranges, but instead would fall down between them. In some places they might fit in the spaces among the oranges, in others they would push the oranges out of place a little. This arrangement in metal is called an *interstitial* (from the word interstice, meaning a

Fig. 6-2. **Alloying atoms in solid solutions may have little effect upon the crystal lattice (a), or may distort the lattice (b), (c), to make the metal stronger and harder.**

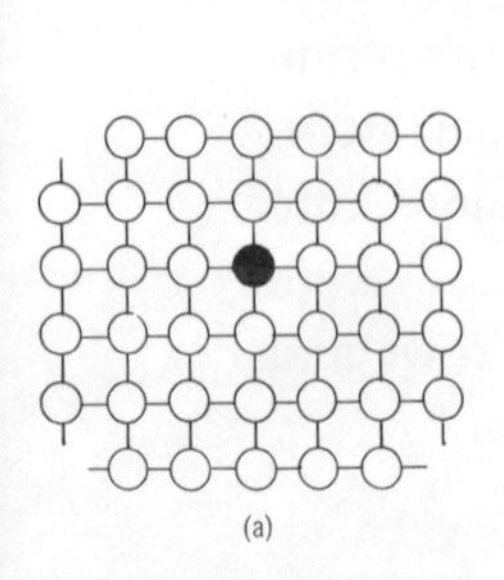

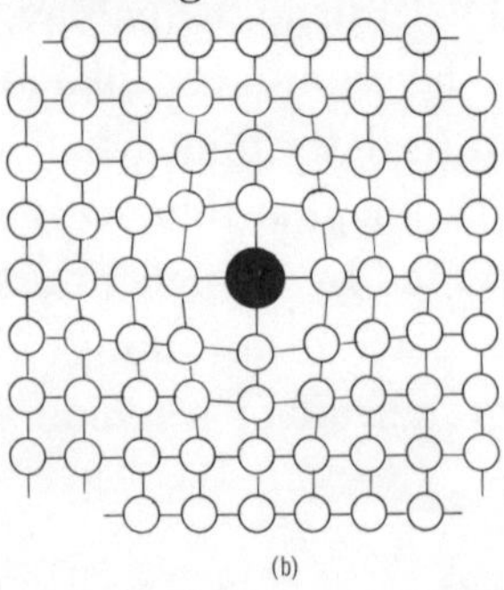

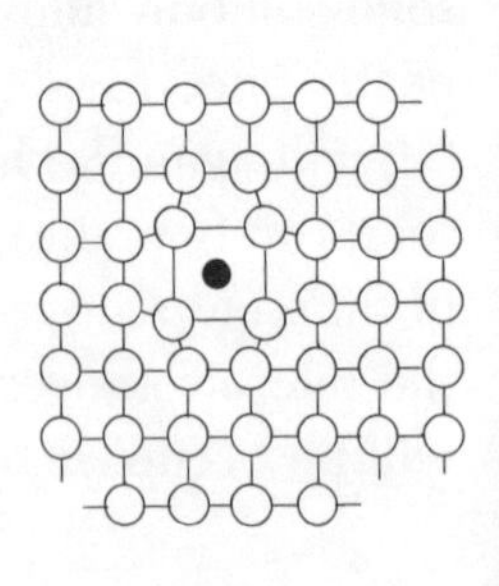

Substitutional alloy

Interstitial alloy

crevice or small space between things or parts) alloy because the smaller atoms are located inside the crystal, among the larger atoms. The cherries would wedge between the oranges if we tried to push a layer to make it slide over the one beneath it. A larger force would likely be needed. The effect of an interstitial atom is shown schematically in Figure 6-2. By distorting the lattice, the added atoms make the alloy stronger and harder than the original metal.

The commonest example of an interstitial alloy is *steel.* Steel is an alloy of carbon (a nonmetal) and iron. The carbon, which may be in the form of carbon compounds, is present in small atoms that lodge themselves among the iron atoms.

In either kind of alloy (solid-solution or interstitial) we can look upon the added atoms as acting somewhat like "pegs," which make it more difficult for slip to occur between the crystal planes in which they are located. Anything that makes slip more difficult makes metal stronger and harder.

The effect which even a small amount of another metal can have upon a soft, weak metal is quite surprising. Copper, for example, is a metal that is unsatisfactory for almost any purpose (except electrical use) as a pure metal. It makes poor castings; it is too soft and tough to machine readily; it is expensive, and does not have particularly good strength. But with small amounts of alloying elements, copper is a wonder metal that can serve a broad spectrum of uses.

The most common alloying elements used with copper are zinc, tin, nickel, silicon, aluminum, and beryllium. Alloys of copper and zinc are commonly known as *brasses.* These alloys can be cast, machined, and worked in almost any way. (Working metals is the topic of Chapter 8.) Because

TYPICAL COPPER ALLOYS

General Use	*Constituents, %*	*Common or trade name*
Lamps, ornaments, hardware	65 Cu, 35 Zn	Common brass
Cartridge cases, tubes	70 Cu, 30 Zn	Cartridge brass
Water pipes, plumbing parts, hardware	85 Cu, 15 Zn	Red brass
Ornaments, plated ware	47–65 Cu, 7–30 Ni, remainder Zn	Nickel silver
Hardware, screen wire	90 Cu, 10 Zn	Commercial bronze
Gears, bushings, bearings	91–97 Cu, 3–9 Sn, 0.5 P	Phosphor bronze
Gears, valves, structural parts, bearings	88 Cu, 8 Sn, 4 Zn	Naval bronze
Fittings, bolts, screws for boats	96 Cu, 3 Si, 1 Mn, trace of Fe	Everdur bronze
Structural parts	89 Cu, 7 Al, 3 Fe	Aluminum bronze

KEY:

Ag	Silver	Li	Lithium	Sb	Antimony
Al	Aluminium	Pb	Lead	Si	Silicon
Bi	Bismuth	S	Sulfur	Sn	Tin
C	Carbon	Mg	Magnesium	Th	Thorium
Cd	Cadmium	Mn	Manganese	Ti	Titanium
Co	Cobalt	Mo	Molybdenum	V	Vanadium
Cr	Chromium	Nb	Niobium	W	Wolfram (Tungsten)
Cu	Copper	Ni	Nickel	Zn	Zinc
Fe	Iron	P	Phosphorus	Zr	Zirconium

***Fig. 6-3.* Constituents and uses of typical copper alloys.**

zinc is cheaper than copper, brass also costs less than copper. The yellow brasses we see in hardware and lipstick cases contain from 20 to about 36 percent zinc. Red brasses are more the color of copper because they contain only 20 percent zinc. Red brass is commonly used where corrosion resistance, workability, or color is important. A list of some copper alloys and their constituents is included in Fig. 6-3.

Alloys of copper that do not contain zinc are generally

called *bronzes.* Bronzes of tin and copper have been known since antiquity and are still used today. Tin increases the hardness of copper and its resistance to wear far more than zinc does. With a small amount of tin (up to 8 percent), bronze is suitable for general use, including coins. With more than 8 percent tin, bronze can be made into strong gears, machine parts, and bearings.

Adding only 10 to 20 atoms of an alloying metal to each 1,000 atoms of copper increases the strength of the copper many times over. Pure copper has a tensile strength of about 5,000 pounds per square inch (psi). That is, a 1-inch-square bar of copper would support a 5,000-pound weight before stretching out of shape. Brass may have a strength 4 times that of copper, or 20,000 pounds per square inch. Special bronzes may have tensile strength 16 or more times that of copper. A 1-inch-square bar of such bronze could support 80,000 pounds.

Aluminum can be added to copper in small amounts, along with certain other elements, to make *aluminum bronzes.* These very strong alloys are used for machine parts and tools, as well as for jewelry. A small amount of silicon (a nonmetal) helps to make a strong, corrosion-resistant bronze, useful for marine parts and fittings. The metal beryllium is a powerful alloying element. A few percent of beryllium in copper makes a strong bronze that is useful for springs, surgical instruments, and watch parts.

Aluminum is another metal that is soft, ductile, and not very strong. It is not very useful in pure form, but is used most often in alloys containing small amounts of copper, silicon, magnesium, or zinc (Fig. 6-4). In addition to increasing its strength, some metals improve the aluminum in other ways. Magnesium improves its machinability, while

General Use	*Approximate Constituents, %*	*Common or Trade Name*
		Aluminum alloys
Structural parts of airplanes, motor vehicles	94 Al, 4 Cu, small amounts of Mn, Fe, Si, Mg	
Lightweight parts and castings	87 Al, 10 Si, small amounts of Fe, Mn, Cu	
Strong; wear-resistant parts	88 Al, 10 Cu, small amounts of Fe, Mg	
		Magnesium alloys
Forgings, castings, rolled products, extrusions products	Mg plus small amounts of such elements as Al, Zn, Mn, Zr, Li, Th, Ag	

***Fig. 6-4.* Constituents and uses of typical aluminum and magnesium alloys.**

manganese and chromium tend to increase corrosion resistance. The addition of silicon makes aluminum easier to cast.

The compositions of some common alloys of several other metals are shown in Fig. 6-5.

Steel, as we have learned earlier, is an alloy of iron and carbon. Other elements can be added to steel to improve even more of its strength, toughness, or hardness (Fig. 6-6). The strength of steel compared with unalloyed iron is quite remarkable. Pure iron has a tensile strength of about 20,000 psi (Fig. 6-7). Plain carbon steel, which contains only one atom of carbon or less for each 100 atoms of iron, may have a strength of 100,000 psi. Alloy steels, containing a small amount of carbon and moderate amounts of other elements such as tungsten, chromium, vanadium, manganese, or molybdenum, may have a strength (when heat-treated) of more than 200,000 pounds per square inch, or more than ten times the strength of pure iron.

Fig. 6-5. **Constituents and uses of some miscellaneous alloys.**

General Use	*Approximate Constituents, %*	*Common or Trade Name*
		Nickel alloys
Hot parts such as exhaust pipes for engines, heating coils, furnace parts	77 Ni, 15 Cr, 7 Fe, small amounts of Mn, Cu, Si, C, S	Inconel
Chemical-handling equipment, food machinery	66 Ni, 29 Cu, 3 Al, small amounts of Fe, C, Mn, Si, S	k monel
High temperature heating elements	80 Ni, 20 Cr	Chromel A
		Titanium alloys
Sheet, plate, wire	Very small amounts of Fe, C remainder Ti	
Bars and forgings, high strength	95 Ti, 3 Cr, 1 Fe, Tiny amount of C	
		Soft-metal alloys
Household utensils, ornaments	85–90 Sn, 10–15 (total) Pb, Cu, Sb	Pewter
Soldering	60 Sn, 40 Pb	Soft solder
Cast printing type	90 Pb, 10 Sn	Type metal
Electric motor bearings	85 Sn, 10 Cu, 5 Al	Bearing metal
Fire sprinkler valves	25 Pb, 12½ Cd, 12½ Sn, 50 Bi	Wood's metal
Die castings	95 Zn,4 Al, 1 Cu, .05 Mg	Die casting metal

Alloying is metallurgy's most important tool for increasing the strength and hardness of metal. The metallurgist can control the strength and hardness of an alloy by heating and cooling the metal in special ways. Ancient metalworkers found by trial-and-error that heating a metal red-hot and then allowing it to cool slowly would make it soft and

Fig. 6-6. **Constituents and uses of typical carbon steels and steel alloys.**

General Use	*Approximate Constituents, %*	*Common or Trade Name*
	(iron is principal component)	Typical carbon steels
Low cost, easy to work, ductile and tough. Auto bodies, containers, wire	.08–.15 C	Low carbon steels
Stronger but less ductile. Buildings, bridges, boilers, trucks, railroad equipment	.15–.25 C	Mild steels
Shafts, machine parts, ship structures	.25–.35 C	Medium carbon steels
Heavy machinery forgings	.35–.65	
Railroad rails and wheels	.65–.85	
Springs, cutting tools (heat treated)	.85–1.2	High carbon steels
		Typical steel alloys
Cutting tools, drills	0.6–1.3 C, 14–20 W, 3–5 Cr, 1–4 V, Mo, Co	High Speed steels
Cutlery, tableware	0.3 C, 12.5 Cr	Cutlery (stainless) steel
Wide general use in cooking utensils, food processing, chemical industry, hospital equipment	0.1 C, 18 Cr, 8 Ni	18/8 Stainless steel
Turbine discs and blades and other parts where heat resistance is important	0.1 C, 10.5 Cr, 6 Co, 1.7 (total) Mo, Nb, V	Gas turbine steel
Structures, machine parts	0.12 C, 6.5 Si, 2.2 (total) Cu, Cr, P, Mn, Ni, Mo	Structural steels Low strength
Heavy disks, gears, axles, connecting rods	0.4 C, .3–.6 Mn, 1.5–2.0 Ni, 0.2–.35 Mo	Medium strength
Heavy duty parts, gears, shafts	.25–.45 C, 0.4–0.7 Mn, 2.3–4.3 Ni, 0.5–1.4 Cr, 0.2–0.7 Mo	High strength

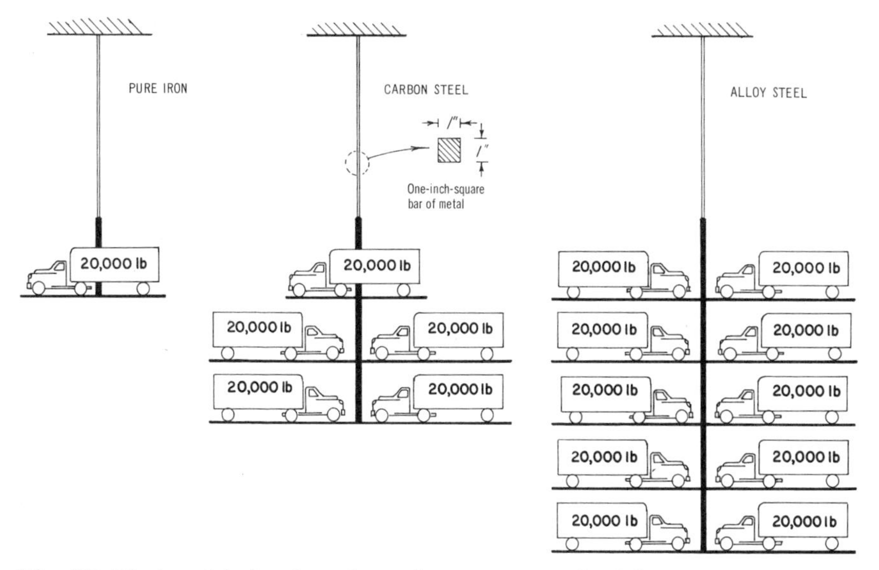

***Fig. 6-7.* Various kinds of steel can have a strength of five to ten times that of iron.**

workable. This is the *annealing* process described in the last chapter. They also learned that cooling their hot steel suddenly by *quenching* it in cold water would make it harder and stronger. This general process we call *heat-treating.* The early metalworkers didn't know what happened inside the steel in the heat-treating process. A metallurgist of the sixteenth century ventured an explanation, perhaps drawn from the effects of hot and cold baths on his own body:

> The long firing (heating) opens up and softens its pores, which are drawn together again tightly by the power of the cold of quenching water. The iron is thus given hardness and the hardness makes it brittle.

Only in this century have we come to understand these ancient processes and learned how to heat-treat all kinds of steels, as well as alloys of other metals, such as aluminum and copper.

Alloys that can be successfully heat-treated are usually those in which the solubility of the alloying metal is limited and decreases as the temperature decreases. This means that the parent metal will dissolve and hold in solid solution a certain amount of the alloying metal at a high temperature. But as the metal cools, the alloying metal is forced out of the crystal lattice and forms crystals of its own. (Remember that although the metal is solid, the atoms can still move around, especially at high temperatures.) When we remember that the presence of atoms of the alloying metal in the crystal lattice increases the strength of the parent metal, we can see that when the alloying metal leaves the solid solution, the metal may lose some of its strength. Because the alloying atoms cannot leave the crystal lattice as easily when the metal is cold, we can force them to stay in solution by plunging the hot metal into cold water to cool it very suddenly.

The heat-treating process is different for each alloy, but the general idea is to heat the metal hot enough and for a long enough time to disperse the alloying element throughout the metal crystals. Then the metal is cooled so rapidly that the alloying atoms do not have time to move out of the crystals, but remain dispersed. In this way, the solid solution, with its desired hardness and toughness, is retained.

Heat-treating steel locks the carbon into the crystal lattice in a similar way. The arrangement of the atoms (and the hardness of the steel) changes very little after the quenching operation. But in some other alloys the atoms can continue to change their positions even after the metal has cooled to room temperature. For example, aluminum alloyed with a small amount of copper can be heat-treated (quenched) to lock the copper into the aluminum lattice. But even after the metal has cooled, the copper atoms continue to move slowly out of their positions in the lattice and clump together as tiny particles of copper. Because these particles tend to "peg" the lattice to prevent slip, the metal continues to harden for a period of several days after the heat-treatment. This process is called "age-hardening." If the alloy was heated again to even a moderate temperature, enough copper might move out of the lattice to form large grains of copper, and the strength of the alloy would decrease.

Many alloys of aluminum can be both heat-treated and age-hardened. Bronzes that contain aluminum and beryllium can be heat-treated. The greatest number of heat-treatable alloys are the steels. A great variety of characteristics—softness, hardness, toughness, shock resistance, etc.—can be produced in various steel alloys by using appropriate temperatures, soaking times, and rates of cooling (quenching). To obtain different rates of quenching, steel can be quenched in cold water, hot water, oil, liquid salt, or even in a liquid metal.

Steel is easily the most important alloy in the history of the world. It was a mystery to the ancients who discovered it, and an object of superstition and worship. As a metal, it was poorly understood until recent times. The story of steel, from the ancient mystery to today's incredible steels, is traced in the next chapter.

CHAPTER 7

INSIDE IRON AND STEEL

A great deal of mystery and superstition surrounded the art of metalworking in the early days. Some people regarded iron with great suspicion, while others thought that iron would protect its owner against evil spirits. In ancient times there were rules in some countries that forbade the use of iron for cutting herbs or slaughtering animals. Some primitive tribes today still believe that iron hoes will keep away rain. Because early people did not understand the process, smelting a bright stream of hot metal from rocks must have seemed to them the greatest of magic. How could anyone get this amazing new material out of useless gritty rocks, except by magic?

But there was a more substantial reason for thinking that metal was hooked up with magic. Things sometimes happened when iron and steel were worked that were very strange and totally unexplainable. When an iron worker found his whole batch of iron suddenly ruined, for no apparent reason, it was easy for him to blame magic or evil spirits.

When the first iron was smelted—and we shall never know just when or how that happened—the metalworker must have discovered that this new stuff was *different.* It came from the furnace as a spongy, porous mass of metal, cinders, and slag. We can be reasonably sure that the first attempt to work this mass into usable metal was a failure. Copper, which had been in use for some 2,000 years when iron was discovered, could be hammered into tools while cold. Iron, on the other hand, shapes easily only while it is red-hot. A hammer will bounce off of cold iron and change its shape hardly at all. We might even guess that the smelter gave up when he found that the new metal—whatever it was—could not be pounded in the usual way. Perhaps iron was actually smelted numerous times over many years, and then abandoned in disgust. But someone eventually hit upon the solution, either by inspiration or by luck. Perhaps he threw the mass back into the fire in a desperate effort to see if more heat would improve it and happened to hit it a lick with his hammer while it was still red-hot. In a shower of sparks, he discovered that iron can be hammered easily when red-hot! He also found that the charcoal and slag squeezes out of the soft, hot mass of metal, leaving nearly pure iron.

The method of hot-working iron was a major discovery, and cleared the way for ironworking to begin in earnest. But iron was not as useful as bronze because it was too soft and weak. A way was needed to make iron harder and stronger, so that it would make good tools and weapons. As it happens, the act of reheating iron again and again to get it ready for each hammering operation can harden it a little at a time. Someone must have noticed this and tried heating solid iron for a long time, buried in red-hot charcoal. To his amazement, the iron that resulted could still be hammered into shape, but turned out to be much harder and stronger when it cooled! If heating in red-hot coals for many hours improved iron, it was natural that metalworkers should try longer and longer periods of heating, plus higher temperatures. Sure enough, they found that the longer the heating period and the hotter the fire, the better the iron—until one day, disaster struck. The pieces of solid iron being heated among the glowing coals suddenly melted down into a blob of liquid metal. When the ironworker removed this mass from the fire and tried to hammer it, the metal shattered into pieces. After the pieces had cooled, he found that the iron was weak and brittle and practically useless.

The mystery deepened when he tried to make this brittle blob of iron useful again. He hammered it, reheated it, broke it, and remelted it, but it did not change. It remained a weak, brittle, unworkable metal. Obviously, the evil spirits had been offended by something and had ruined his metal. He did what he had to—threw it away and wearily started over with fresh iron ore and charcoal to make a new batch of iron. We know that this minor disaster happened over and over again, because many discarded chunks of brittle iron have been found near ancient ironworks. Not for several thousand years did anyone discover what made iron brittle and how brittle iron could be changed back into the useful kind.

What had happened to the iron to first make it soft and

malleable, then hard and strong, then suddenly weak and brittle, is quite simple. Iron can absorb the element carbon, and even a small amount of carbon dissolved in iron greatly changes the nature of the metal (Fig. 7-1). The first pasty mass that the metalworkers smelted from iron ore had almost no carbon in it—because it had not been heated to a high enough temperature or for a long-enough time to absorb an important amount of carbon from the red-hot coals in the fire. When the iron was heated red-hot again and again for more hammering, it absorbed a very small amount of carbon each time, depending upon how long the ironworker left the piece buried in the fire. Through this kind of experience, metalworkers learned to "soak" the iron in red-hot coals for hours or even days. When this was done, the iron slowly absorbed carbon on its surface, giving the iron a thin, hard shell. This shell was actually the alloy of iron and carbon we call *steel.* The process is known as *carburizing* and is still used today.

The longer the solid iron is soaked in red-hot coals and the hotter the fire, the faster and deeper the carbon is absorbed. If the temperature of the fire is raised to 200° above the smelting temperature, the iron begins to absorb carbon very rapidly—almost like a blotter absorbs ink. Because carbon lowers the melting point of iron, the pasty mass of iron will suddenly melt and run together. By this time it will have absorbed nearly as much carbon as it can hold—perhaps as much as 4 or 5 parts of carbon for each 100 parts of iron. This alloy is not as strong as either wrought iron or steel, but is brittle and weak. It cannot be hammered into another shape, and therefore is not useful for making tools that must be tough and strong or hold a sharp edge. It can only be poured into molds to make heavy machine parts

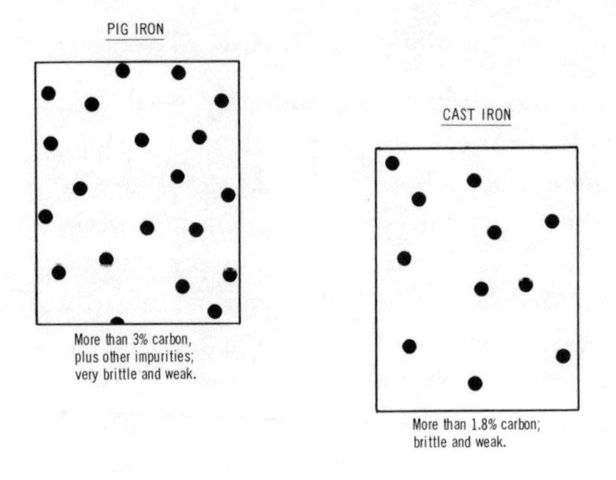

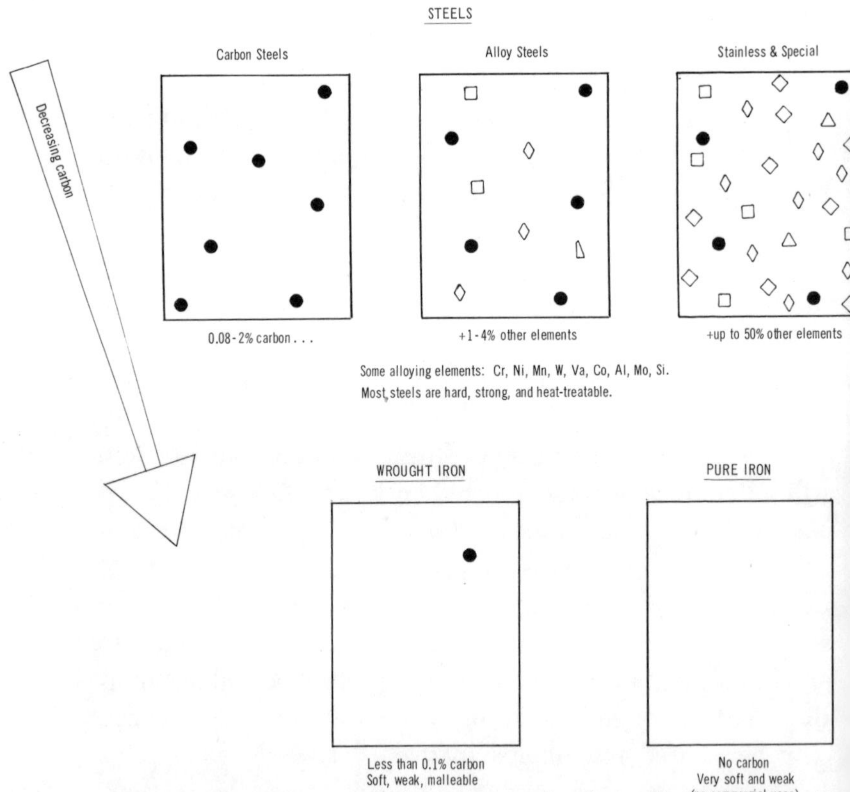

Fig. 7-1. **Small amounts of carbon and other alloying elements can greatly change the characteristics of iron.**

for which strength is not important. We call it "cast iron," and we use a great deal of it today. The ancients, with their need for small hardware and sharp tools, could find no use for it at all.

Even if they had known that the cause of their brittle iron was too much carbon, they had no way of getting the excess carbon out. Heating, melting, or hammering the iron will not make it give up the carbon it has absorbed. So metalworkers had to learn not to go too far in the tricky business of changing crude iron into steel. They had no way of measuring the temperature of their metal, but learned to control the temperature by watching the color of the glowing iron as it went from dull red, to bright red, to orange, to yellow, to dazzling white (Fig. 7-2). They had no way of knowing what was in the steel they made, but they learned to judge its quality by the way it felt under the hammer. Because of this need to rely on experience and "feel," early metalworking was more an art than a science. The skills of working iron—and all metals—were passed down from father to son, from worker to apprentice, for dozens of generations.

By trial and error, men who worked the new steel discovered that the manner in which the metal was cooled seemed to make a difference in its hardness. When red-hot steel is allowed to cool in air it is tougher and harder than iron. But if it is cooled very quickly by quenching in cold water, it becomes still stronger and harder. Steel heat-treated in this way can be made so hard that it can be used to make tools for cutting other softer metal.

Superstitition led to a strange and cruel method for quenching steel that was believed to give a superior sword blade for use in war. A prisoner of war or a slave was used instead of a tub of water. With the unfortunate fellow held or

Color	Approximate Temperatures Degree C	Degree F
First visible color, black red	525	970
Dull red	700	1300
Blood red	800	1500
Cherry, bright red	900	1650
Bright cherry	1000	1800
Dull orange	1100	2000
Bright orange	1200	2200
White	1300	2400
Bright white	1400	2550
Dazzling white	1500	2700

***Fig. 7-2.* The color of glowing metal.**

tied in the proper position, the sword was heated red-hot in the fire and plunged downward through his body. The blade cooled in an instant and was withdrawn and quickly swung in a powerful motion to cut off the slave's head. If the sword cleanly severed the head from the body, it was considered a good blade, ready for the hand of a soldier. If it did not, the blade was put back into the fire and another slave brought in to try again.

Steel that has been hardened by quenching tends to be brittle and may break easily, especially if given a sharp blow. Early metalworkers discovered that the steel could be improved still more by heating the metal once again to a temperature well below the temperature from which it was quenched and allowing it to cool slowly. This process, called *tempering*, made the metal tougher and less brittle without reducing the hardness and strength by very much. Again, the metalworker had to judge the temperature by watching the color of the glowing metal and using his hard-bought experience to get the hardness and strength he wanted.

When metalworkers learned to carburize, quench, and temper steel, the age of iron and steel came into full flower. By means of these processes, learned by trial and error and passed down from father to son, metalworkers were making fine steel between two and three thousand years ago. Samples of steel dating from over 2,500 years ago (650 B.C.) have been analyzed and found to have few impurities and to have the amount of carbon required for high quality steel.

Although they made good steel, ancient metalworkers never solved the problem of changing brittle iron back into useful iron or steel by removing its excess carbon.

This problem remained an obstacle to the large-scale production of iron and steel for a very long time. Iron could be smelted more easily and in larger amounts by using high temperatures and a blast furnace, as we do today. Unfortunately, the high temperatures needed to smelt iron into a free-running liquid will also cause the iron to absorb too much carbon. Such iron (which is called pig iron) could be used for making cast items where brittleness does not matter. But no one knew how to make pig iron into soft iron or steel.

A breakthrough on this problem came only about 400 years ago when the *finery* hearth was developed. In the finery method, bars of brittle pig iron were heated in a charcoal forge to near melting temperature. Then a blast of air from a bellows was directed at the end of the red-hot iron. The air burned some of the carbon in the iron, releasing heat as it did so. The extra heat melted the iron from the bar in droplets, which dripped down to solidify into a mass of soft iron that contained very little carbon. This mass could then be hammered and worked into any shape desired, or processed into steel by carefully adding back a tiny amount of carbon. Obviously, this process was slow and expensive.

Later, during the early 1700s, the French inventor, René Réaumur (1683–1757), developed another method for getting the excess carbon out of brittle iron. He buried bars of pig iron in powdered iron ore and heated them to a bright red color for many days. You can probably guess what happened in this process: at high temperatures the oxygen in the iron ore is more strongly attracted to carbon in the pig iron than to the iron in the ore. The oxygen therefore pulled the carbon out of the pig iron, converting the metal back into low-carbon wrought iron. This too was a slow and expensive process.

Réaumur did not know exactly why his process worked. However, he was very close to the truth when he speculated that wrought iron was pure iron and that it became steel when some other substances were added. He called these unknown substances "sulfur and salts." It was not until more than sixty-five years later that the sciences of chemistry and metallurgy proved that his mysterious "sulfur and salts" was really the element *carbon*.

One of the processes used to make steel from the wrought iron made in the finery or by the Réaumur method was to heat soft (nearly pure) iron in sealed containers, or crucibles, with a small amount of carbon. When heated red-hot for many hours, even days, the iron would eventually absorb just enough carbon to melt down into a small cake of high-carbon steel. This process—called the *crucible process*—made excellent steel, but it was slow and not suited to mass production.

What was needed, now that carbon had been identified as the mysterious ingredient in iron and steel, was a fast, inexpensive method for removing carbon from the pig iron that could be so easily produced in the blast furnace. Because the demand for steel was growing rapidly, a fast, cheap process for making steel was also needed.

Two iron experts, one in England and one in America, independently hit upon an idea in the mid-1800s that changed the whole iron industry. At that time, about two weeks was required to make 40 or 50 pounds of steel from soft iron by the crucible method. With the new process that these two men developed, as much as 10,000 pounds of steel could be made from brittle pig iron in 20 or 30 minutes.

The method was basically the same as that used in the finery: burn the carbon out of pig iron using the oxygen in air. The new twist was to use a large container of *molten* iron. Sir Henry Bessemer (1813–1898), the Englishman, began his experiments with a kind of open-hearth furnace in which air and the flames from burning fuel are drawn over a bath of molten metal. Then he hit upon a new idea. Instead of trying to burn out the carbon by blowing flames and air across the surface of the molten metal, why not bubble the air up *through* the molten metal? It was a daring new idea.

He built an egg-shaped vessel he called a *converter* and piped in compressed air through the bottom. When he filled his converter with molten iron and blew air up through it, the converter erupted like a volcano, shooting sparks, flames, and smoke high into the air (Fig. 7-3). When the flames died down after ten minutes or so, the metal was tested and found to be soft and malleable. The flames and sparks had been produced by the burning of the unwanted carbon and other impurities.

Although Bessemer's converter showed how excess car-

***Fig. 7-3.* A *Bessemer converter* during a "blow." (*American Iron & Steel Institute*)**

bon could be removed from iron, people who tried to use his process to make steel ran into trouble right away. There still remained other impurities in most iron that were not being removed by the converter. Also, a small amount of carbon had to be left in the iron (or put back) to make it into steel. These refinements to the basic process proved difficult to accomplish, but were eventually solved by a brilliant metallurgist, Robert Mushet (1811–1891). Mushet recognized that Bessemer's converter was not only burning away the carbon but also putting so much oxygen into the iron that it was "burning" the iron as well. He added a compound of iron, carbon, and manganese to the molten iron. The manganese drew the excess oxygen from the iron, while the carbon put back the tiny amount of carbon the iron needed to become steel. With this process he produced excellent steel.

About eight years before Bessemer's experiments, an American ironmaker in Kentucky named William Kelly (1811–1888), was watching pig iron being refined into drops of wrought iron by means of a blast of air in his finery. He had been puzzled by the fact that the iron under the air blast looked *hotter* than the rest. He suddenly realized that the blast of cold air was actually producing *heat* when it struck the iron, because of the heat released in burning the carbon in these drops of iron. From this discovery, he too got the idea that molten pig iron might be refined into wrought iron and steel by directing an air blast through it.

The idea was so revolutionary that Kelly's family feared for his sanity and had him examined by a physician. The good doctor understood the process when Kelly explained it to him, and he became one of Kelly's strongest supporters. After many years of experiments and numerous failures, Kelly finally succeeded in producing soft steel in his converter at about the same time as did Bessemer. In 1857, Kelly was granted a patent on his process in the United States.

As it turned out, both Bessemer and Kelly needed the Mushet process for controlling the carbon in steel, while Kelly needed the ingenious converter and other machinery designed by Bessemer. The patents of all three men, plus the efforts of other skilled steel men, were finally used by the steel industry to build the converters and other equipment for the mass production of steel. Production of steel in the United States soon rose to 3,000 tons per year; 32 years later (1889), annual production of steel reached 8.5 million tons. Today it is more than 150 million tons annually.

Although the Bessemer converter was the first method to be developed for making steel on a large scale, it was not the last. Within a few years after the Bessemer process was perfected, the *open-hearth* process was developed, at about the same time in both France and Great Britain. The open hearth furnace is a large, rectangular shallow basin made of firebrick, with a domed roof overhead (Fig. 7-4). The charge of molten metal, scrap iron, limestone (flux), and iron ore is

Fig. 7-4. **Molten iron being poured into *open-hearth furnace.* (*American Iron & Steel Institute*)**

Fig. 7-4a. **Schematic view of an open-hearth furnace. Fuel burned in preheated air heats the charge of metal. Iron ore, extra air, and oxygen may be used to remove carbon to the desired level.**

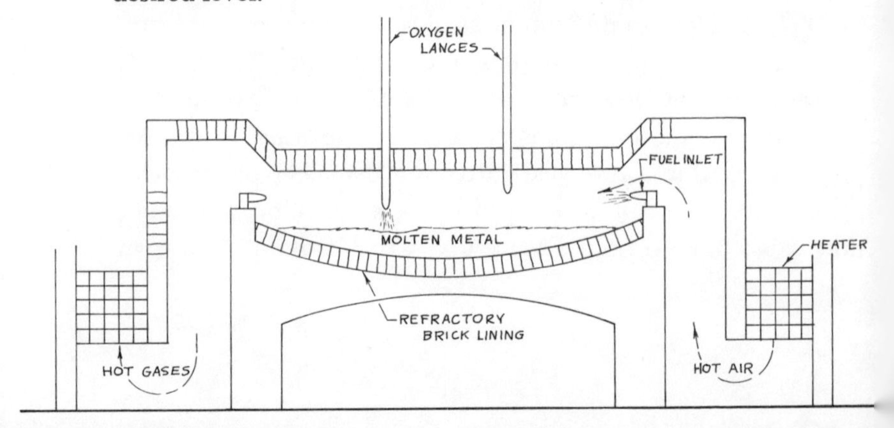

Fig. 7-5. **A charge of molten metal is being poured into this *basic oxygen furnace* for processing into steel. (*American Iron & Steel Institute.*)**

Fig. 7-5a. **Schematic view of a basic oxygen furnace. Furnace is charged with molten iron and an oxygen lance used to burn off carbon and other impurities to the desired level.**

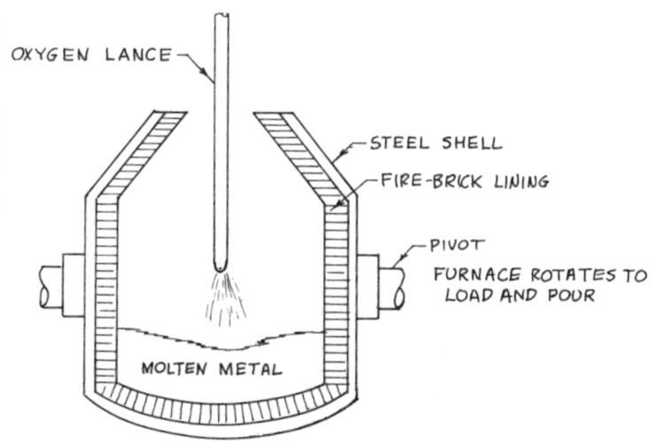

heated by a gas or oil flame over the basin. Various minerals are added to the bath of metal to carry off undesirable substances such as sulfur and phosphorus. The oxygen forced from the iron ore and from the rust in the scrap iron by the intense heat bubbles up through the molten iron to burn out the carbon. In some modern open-hearth furnaces, a blast of pure oxygen from an overhead *lance* (a pipe) is used to hasten the burning off of the carbon and other impurities. Samples of the metal are drawn off periodically and analyzed for the amount of impurities and carbon. Although the open-hearth process is slow—seven to nine hours for one

Fig. 7-6. An electric furnace. (*American Iron & Steel Institute*)

Fig. 7-6a. **Schematic view of an electric furnace. A heavy current of electricity flowing between electrodes and metal melts the charge. Flux and oxygen are used to remove impurities. Alloying elements can then be added.**

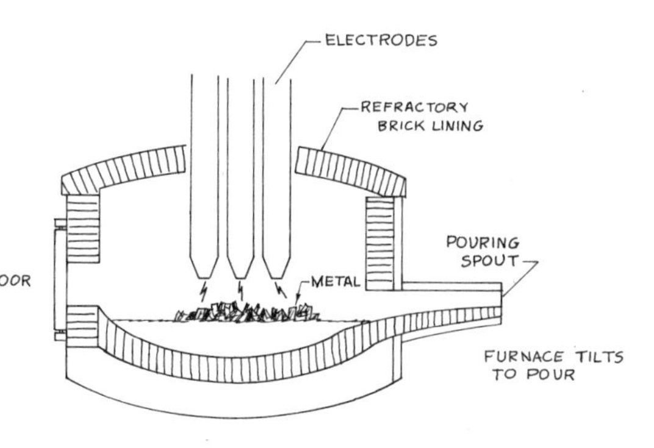

batch—it can make a large batch of high-quality steel of a very exact composition.

The open-hearth furnace became more and more popular, and in 1908 it passed the Bessemer converter in production. Today there are few Bessemer converters in use anywhere. But even the open-hearth furnace is now rapidly being replaced by new developments such as the basic oxygen furnace and the electric furnace (Fig. 7-6).

The mass-production of steel made possible the amazing industrial progress of the past 150 years. From the huge furnaces of the steel industry came materials to build bridges, railroads, ships, tall buildings, motor vehicles, power plants, airplanes, and the factories and machines of all the other industries. One of the greatest developments of this period was the metals industry itself, which had to learn not only how to smelt metal from ores, but also how to shape and work metal into thousands of useful things. Some of the processes that take place inside a metals factory are the topic of the next chapter.

CHAPTER 8

WORKING AND SHAPING METALS

The world's first steam engine—actually a steam-powered pump—was built and put into service around 1700. It was not until about seventy-five years later, during the American Revolution, that James Watt improved on the steam-pump idea and built the first true steam engine in Great Britain. Several more decades were to pass before the steam engine was used to power boats and locomotives.

One of the problems that slowed progress and plagued the inventors of that day was the need for better materials for their inventions. Wood was the standard material for wagons and ships, and wooden shafts and gears had even been used successfully in windmills. But wood was inadequate for building steam engines, most kinds of farm or factory machinery, or railroads. For the big, heavy, powerful machines the inventors were building, metal was needed—good metal, too, and big pieces of it. Working this metal and building these machines required much more skill than was available in the small shops where small tools, hinges, ship fittings, and household equipment were made with hammer and forge. Much new technology and many new skills were needed, not only to make very large pieces of metal, but also to shape them accurately into pistons, shafts, bearings, and frames for the new machines.

James Watt's most difficult problem 200 years ago was not designing his new steam engine, but getting it *built.* The job took years, and the methods for building such a large, new machine had to be invented right along with the engine itself. The workmen who put it together were proud that they had built the piston to fit the cylinder so perfectly that a shilling (an English coin) could not be slipped into the gap between the two. While this may have been a remarkable fit for that day, Watt's engine must have leaked a great deal of steam through that gap between piston and cylinder.

Watt was wary of steam pressure and used pressures of only a few pounds per square inch in his steam boilers. Other more daring inventors used much higher steam pressures. Although their engines produced more power, their boilers also exploded more often. The great pressure of the steam could not be held by the crude rivets and bolts that fastened the boilers together. Even the metal from which the fasteners and boilerplates were made was unreliable.

During this period of invention and industrial development, known as the Industrial Revolution, the inventors of new machinery and the sciences of metallurgy and metalworking made progress together, each creating problems for the others, but also each helping to solve the other's problems. For example, the inventor often needed large parts (like Watt's huge piston and cylinder) made of a better metal than was then available. The metallurgist and metalworker (they might have been the same person in those days) worked to provide better metal and to develop new processes for casting and machining the engine parts. These new processes made possible still better designs or even a whole new invention, which called for still better metal, and so on.

The railroads were held back for a time due to the lack of a suitable rail. Early railroad tracks were made by fastening a strap of iron to a wooden rail. However, such tracks could not carry much of a load. After the steel industry found a way to make strong, heavy, steel rails, the railroads were then able to build better roadbeds, use larger locomotives, and carry more freight. As a result, the railroads became one of the largest users of steel in the country, while the steel industry became one of the railroad's best customers for shipping steel products around the country.

Through this kind of cooperation with the other industries, the metals industry has made great progress over the past several hundred years. Metalworking progressed from a few simple hammering, cutting, and casting operations to a large number of processes, including forging, rolling, stamping, welding, machining, extruding, and others. These processes make possible today the fabrication of everything from locomotives to automatic pencils, from spacecraft to paperclips.

When we speak of *working* a metal, we mean changing its shape permanently in some way. As we learned in earlier chapters, when metal is permanently deformed, the crystals of the metal slip in thousands of places. When such deformations occur at room temperature (or, more exactly, at a temperature below that at which the metal can rearrange its crystal structure), the process is called cold-working. Each slip distorts the crystal and pushes the rows of unit cells up against the crystal boundaries, thereby making the metal stronger and harder. But a metal will not go on getting stronger indefinitely. If it is worked too much, it will reach the point where the crystals will not slip further, but will break instead.

Before this point is reached, the metal can be made soft again by the *annealing* process described in Chapter 5. When the metal is heated to between ⅓ and ½ the melting temperature, the atoms in the crystal lattice can jump to new positions to correct dislocations and to relieve the distortions and strains at the grain boundaries that cold-working has produced. In this adjustment, the metal may "recrystallize." That is, the crystals that have been squeezed and hardened into smaller crystals in the working process may realign their atomic lattices to merge once more into fewer, larger crystals. With larger crystals and a smaller area of grain boundaries, the metal becomes soft and ductile once again.

Often it is desirable to soften a very hard metal without allowing its grain to become too coarse, or to get rid of only the worst strains while leaving the metal quite hard and strong. By carefully controlling the annealing temperature and time, and the rate of cooling of the metal, it is possible to get the desired conditions in the annealed metal.

When early metalworkers found that iron could be worked more easily red-hot than cold, they had discovered the process of "hot-working," which is still used today. Hot-working can be described as a process of simultaneous cold-working and annealing. When metal is worked at a temperature above the temperature at which the metal can recrystallize, the

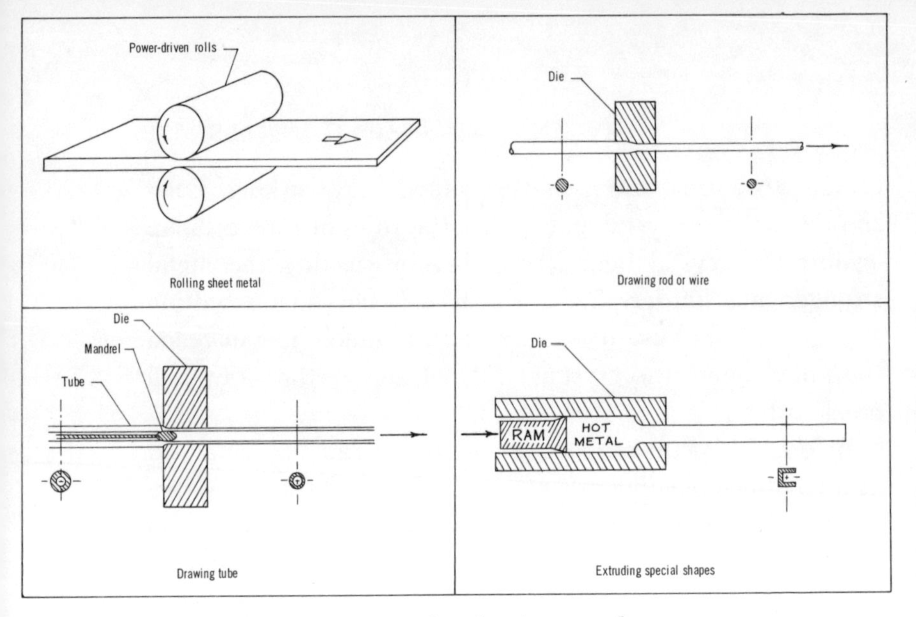

Fig. 8-1. **Four processes for shaping metal.**

strains produced in the structure of the metal are relieved almost as soon as they are produced, because the atoms in the crystal lattice can readjust themselves immediately. The metal never gets hard, but can be hot-worked continuously, without the need for stopping to soften it by annealing.

A few metals—tin, zinc, and lead, for example—recrystallize at room temperature and do not need to be annealed when worked at this temperature. A piece of lead can be bent back and forth repeatedly without breaking. (The metal lead should not be confused with "pencil lead," which is actually graphite, a form of carbon. Graphite is a nonmetal and is very brittle.) The crystal planes in the lead slip, but the atoms can move to relieve the strain at the crystal boundaries immediately, and the metal does not become hard. Lead, in effect, is being hot-worked at room temperature. On the other hand, if we bend an iron coat-hanger back and forth, it quickly becomes stiff and hard and will break if we continue to work it. Iron can recrystallize only at a high temperature.

Early metalworkers also discovered the process of *casting*, which is still a key process today. They poured liquid metal into a mold or a sand bed that was shaped like the object they wanted to produce. When the metal hardened, the mold was broken away, and the casting cleaned and trimmed.

These four processes, discovered in ancient times—casting, cold-working, hot-working, and annealing—are still used today throughout the metals industry. Many of today's newer ways of working metals are only refinements of these very old processes.

One of the first needs of the growing metalworking industry was for a way of producing sheets of metal. This was done in the early days by joining two or more small pieces by hammering them together while red-hot. Then by hand, and later by machinery, they hammered the red-hot plate down to the thickness they wanted. This process was crude and slow and could only produce relatively small pieces of sheet metal.

Fig. 8-2. A plate of red-hot steel emerges from a rolling mill in a steel plant. (*American Iron & Steel Institute*)

Fig. 8-2a. **Aluminum rolled down to foil less than 1/1000 inch thick is a common household item.**

To answer this need the process of *rolling* metal was developed. By passing the metal again and again between a pair of heavy metal rollers, the metal could be squeezed down to a thinner and thinner sheet. This process has been improved over the years to where metal can be rolled into a variety of products from thick plates and giant girders for use in building bridges to paper-thin metal foil.

Some metals, such as brass, are usually rolled in a cold condition and annealed in a separate operation whenever softening is needed. Other metals, such as steel, are usually worked in a hot condition, so that no separate annealing operation is needed. Rolled sheet metal is the raw material from which many products are made.

Metal wire, too, was forged by hand in early days. A piece of metal was cut into a long thin strip and hammered into a rounded shape. Only short, crude pieces of wire could be made by this process. The process of *drawing* wire came much later. When a piece of wire is pulled through an opening, or die, that is a little smaller than the wire, the wire is squeezed down to a smaller size. By passing the wire through one die after another, the wire can be drawn to a diameter smaller than the hair on your head. Figure 8-1 shows how these processes work, while Figs. 8-2 to 8-4 are photographs of factory machinery.

Tubing can also be shaped in a drawing process. A thick piece of metal with a hole in it is pulled through one die after another until the desired size is reached. A center plug in the die shapes the inside of the tube to the desired diameter at the same time the outside of the tube is being shaped. In this way, tubing can be drawn down to very small sizes. The "needle" end of a hypodermic syringe that is used to give flu shots and other injections is a very fine tube drawn down in dozens of steps from a large piece of metal.

Fig. 8-3. A wire-drawing machine in a steel mill. See also *Fig. 8-1.*

Fig. 8-4. Three pieces of copper tubing are being drawn down to a smaller size at one time on this machine. The die is just above the workman's right hand. See also *Fig. 8-1.* (*Anaconda Brass Division*)

In all these processes, the metal deforms, or flows, by slipping within its crystals, as described earlier, and becomes harder and stronger. Sometimes the rolling or drawing operation can be designed to give exactly the degree of hardness desired when the correct size is reached. In such cases, the metal can be left in the hard condition. Often, however, the metal will become too hard to be worked all the way to the correct size. The process must then be interrupted to send the metal through an annealing furnace, in order to soften it. Then it can be sent back for further rolling or drawing.

All of these processes require a great deal of energy. The first rolling operations were done with small rolls turned by muscle power. The first wire was made by pulling it through the die by hand. Only small amounts of material could be produced by muscle power. A great deal more than muscle power was needed for the heavy job of metalworking. In Colonial days, factories were built beside rivers in New England so that waterwheels could be used to turn their metal-working machinery. Today, electric motors drive the machinery of metalworking factories, which are now scattered over the whole country. The large amount of energy used in producing and working metals makes the metals industry partly responsible for our current problems of energy and pollution, as we shall see in Chapter 12.

Another important process which is used to produce machine parts is *forging*. A chunk of hot metal is placed on a platform and pounded (or "forged") into the desired shape by a heavy power hammer that can strike a blow of many tons at each whack. The hot metal, in a step-by-step process, may be forced by the blow into a series of molds that shape it into the part desired. This kind of forging is used to pro-

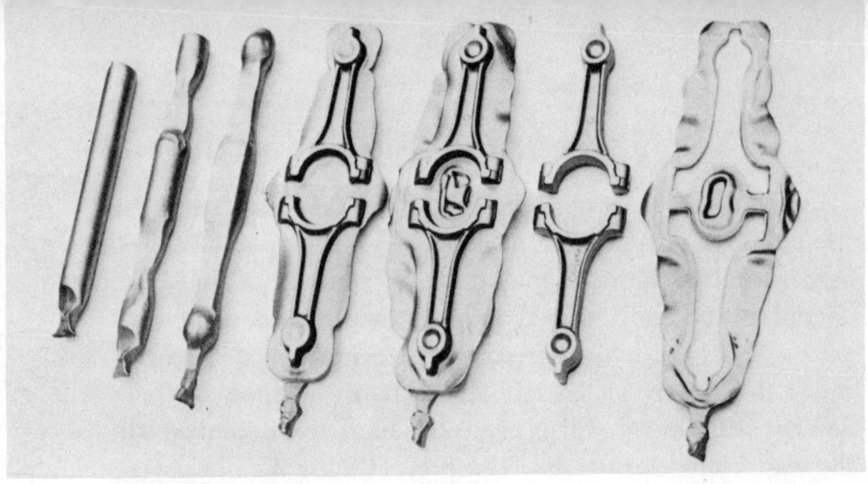

***Fig. 8-5.* The bar of metal at left is forged and trimmed into an engine connecting rod blank, ready for final machining, in five separate steps.**

duce engine parts, such as connecting rods, that need to be very tough and strong. The forging process makes stronger and safer parts than the casting process. Castings may have holes, impurities, or poor crystal structure because of uneven cooling. When a cast *billet* (an ingot or bar of metal) is forged while red-hot, the combination of the heat and the hammering action refines the crystal structure, makes the metal uniform, and welds up any cracks and faults in the raw metal.

A great many of the metal things we use are made by forcing sheet metal into a form, or die, on a powerful press. Automobile bodies, refrigerators, and wheelbarrow beds are made in this way (Fig. 8-6). Sometimes the sheet metal is pulled, or stretched, over a form to make a curved part. Metal canoes and many parts for airplanes can be made by stretch-forming. Or, metal can be pushed down into a deep die to make parts like rifle cartridges, lipstick cases, and pen caps (Fig. 8-7). Several different steps may be required, in which case, the metal is gradually pulled and stretched into the shape desired.

Fig. 8-6. The press at left bends a straight metal bar into an automobile bumper with one stroke. (*Reynolds Metals Company*)

Fig. 8-7. **An automatic machine for deep-drawing metal items. Note strip of brass feeding into machine at left.**

Sheet metal can be struck with a die that raises a design on the surface. The coins in your pocket are made this way. The machine stamps both faces on the coin and cuts it out of the sheet of metal at one crushing blow. Under enormous pressure, the solid metal "flows" into all the hollows of the die to make the design and the lettering on the coin. Tableware (knives, forks, and spoons) are stamped out of metal in much the same way. The machine bends the metal, changes its thickness, raises a pattern on it, and cuts it to size, all in a few quick operations. Because the metal has been "worked" in the stamping or forming operation, it is harder and stronger than the original metal. Items like coins and tableware can be made better and more cheaply by stamping than they could be made by casting or by any other process.

Metals can be joined together in ways that are not possible for other materials. Wood, for example, is most often nailed, bolted, or glued together. Metal can be joined in

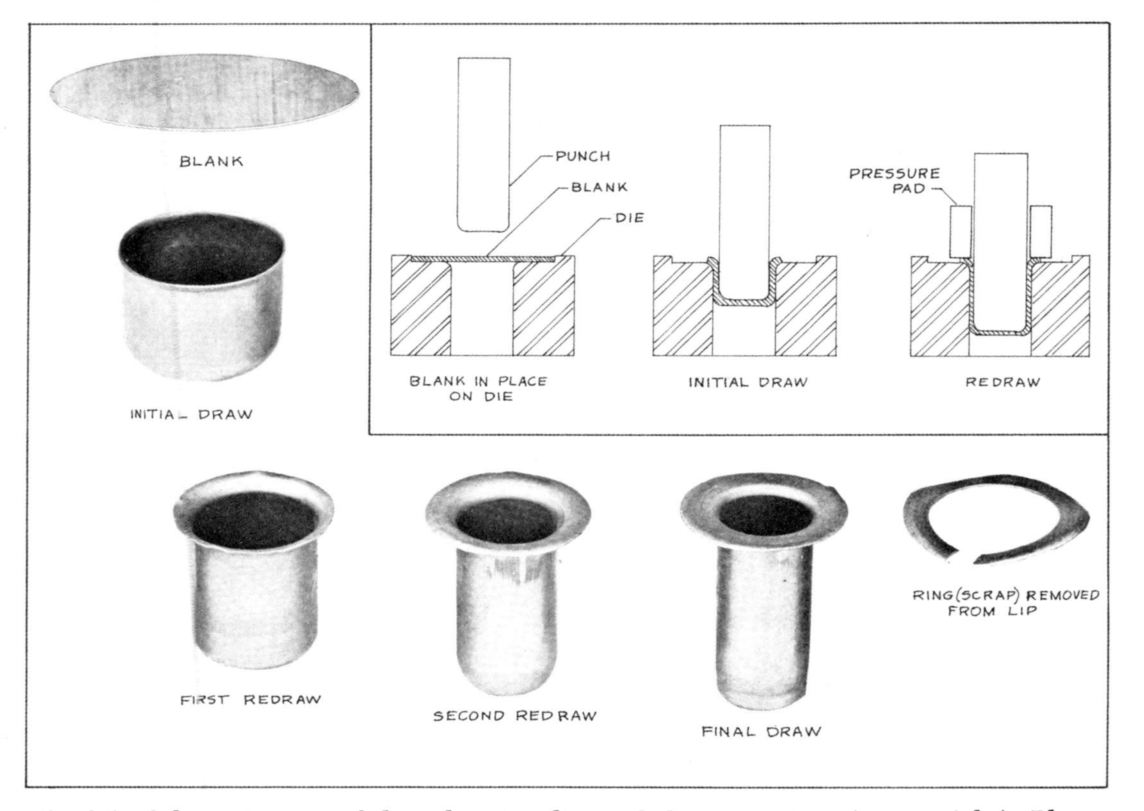

Fig. 8-7a. **Schematic view of deep-drawing dies and their operation (upper right). Photos show various steps in the manufacture of a deep-drawn item.**

similar ways, by bolting, by riveting, or even by gluing. But metal can also be welded—the two parts melted at the joining edge, and fused together, to make a continuous bond of metal.

A kind of welding was used by ancient metalworkers, probably in the very early days of ironworking. They heated two pieces of metal red-hot and then forged them together by hammering. The hot metal was squeezed together so tightly that the two pieces formed a metallic bond. While this was a useful process for small objects, it was not easily used for large pieces of metal.

About 100 years ago, new ways of welding metal together were developed in which the edges to be joined are actually melted and fused together to make a very strong joint. The metal can be melted by a gas torch or an electric arc, while, at the same time, new metal from a welding rod is melted into the joint to complete the bond. In some newer processes the melting is done by an electron beam or a laser beam.

Welding has led to great improvements in metal fabrication. In a ship's hull the plates are welded all along their edges to give a smooth, strong, watertight joint. In automobile bodies, the two parts may be joined by a process called *spot-welding*. The two parts are clamped tightly together while a large current of electricity is passed through them for an instant. The current melts the metal and fuses the two parts together at that point. By moving the welding tool from one point to another, the two parts can be "stitched" together by a row of spot-welds.

Some metals, such as steel, are easily welded. Others, like cast iron and aluminum, are more difficult. Special care must be taken to keep the air from oxidizing the molten metal in the weld and spoiling the bond. For most metals, a "flux" is used to reduce the oxidizing and to carry off the impurities. For some more active metals, such as aluminum, even greater protection may be needed in the form of a blanket of an inert gas, such as helium. This is flooded over the welding process to keep the air from reacting with the melted metal.

We might wonder whether two pieces of clean, cold metal that are just squeezed together very tightly might join in a metallic bond. The answer is that such a joint usually would not be "perfect" enough to allow the atoms to get close enough to each other to bond. The surfaces are likely to be at least slightly corroded, irregular, or coated with oil and gases from the atmosphere. Under some conditions, however, some metals can be welded without heat if they are well-cleaned and squeezed together with an enormous force. This process is called *pressure welding.*

With the coming of space travel, metallurgists began to wonder whether the vacuum of space might clean the metal so well of all air and other gases and even of lubricants that two metal parts might weld together accidentally if the two were rubbed or clamped together. Experiments showed that such welding might indeed occur and cause working parts of spacecraft to "freeze" and stick together. Special coatings and lubricants are used on spacecraft to prevent such disasters from happening out in space.

Two other methods of joining metal, *soldering* and *brazing*, were also used in ancient times. Examples of both have been found from as far back as early Roman times. In the process of soldering, a soft alloy of tin and lead, called solder, that melts at a low temperature is allowed to flow between the clean surfaces of two pieces of metal to be joined. The solder bonds to both pieces with a metallic bond and cools to form a fairly strong solid-metal joint between them. This process can be used only for certain metals and is used mostly to join small parts and to make electrical connections. The brazing process is similar to soldering, but uses stronger bonding metals and higher temperatures. Neither process melts the metal to be joined—only the bonding metal is melted. Both soldering and brazing are still widely used today.

A manufacturing process that illustrates most dramatically the great plasticity of metal is *extrusion*. In this process, hot metal is forced, under enormous pressure, through a die. Although solid, the metal flows through the die like toothpaste from a tube, forming a rod, bar, tube, or any shape for which the die has been designed. Extrusion can be used to produce complicated shapes that would be difficult to make in any other way (Fig. 8-8). Most aluminum windows and doors are made from extruded shapes. Extrusion works best with soft metals, such as aluminum or magnesium.

Besides the shaping operations described above, metal parts can be shaped by cutting away excess material. Changing the shape, size, or surface condition of a part by cutting away material is called *machining*.

It has been estimated that 15 million tons of metal are whittled away into chips every year in the United States, at a cost of perhaps 34 billion dollars! Most of this machining is done by pressing a sharp tool against moving metal and straining the material until it breaks away in chips or long streamers. Several of the simpler processes, namely turning, planing, and milling, are shown schematically in Fig. 8-9. As the sharp-edged tool plows through the metal in any of these processes, the metal crystals slip over each other, like a deck of cards, and chips or streamers of metal are peeled away from the surface of the metal.

Obviously, the cutting tool must be made of harder and stronger metal than the metal it must cut. Soft metals can be cut with tools of carbon steel. Hard metals, such as steel itself, may require extremely hard tools made of an alloy steel containing metals such as tungsten, chromium, molybdenum, vanadium, and cobalt. Because the tool is made hot by friction as it cuts away metal, the tool material used must have great strength and hardness at high temperatures. Sometimes a cooling fluid, such as oil, is squirted on the work at the point of cutting to keep things cool.

The smoothness of the metal after it has been machined depends upon many factors, including the kind and hardness of the metal, the depth of cut, speed of cutting, and the kind of cutting tool used. A deep, rough cut can be fol-

Fig. 8-8. These complicated aluminum shapes, produced by extrusion (see _Fig. 8-1_), would be difficult to manufacture any other way. (_The Aluminum Association_)

lowed by a fine finishing cut that will leave the metal smooth. If a still smoother finish is desired, the surface may be ground by a fine-grain abrasive grinding wheel (Fig. 8-9).

The ability to machine metal parts accurately has made possible mass production and interchangeable parts. Until this idea was introduced in the United States, each part of a device, such as a gun, was made separately by hand and would fit only that gun. Eli Whitney, who is best known as the inventor of the cotton gin, was also a famous manufacturer. His factory made the first use of mass production in the U.S. when he produced 25,000 muskets for the government, using interchangeable parts. Today nearly everything is manufactured in this way.

If James Watt were to build his steam engine today, he would find that it need not be built of crude castings, machined to fit "within the thickness of a shilling," but could instead be forged, welded, and machined to clearances of a

***Fig. 8-9.* Four processes for cutting metal.**

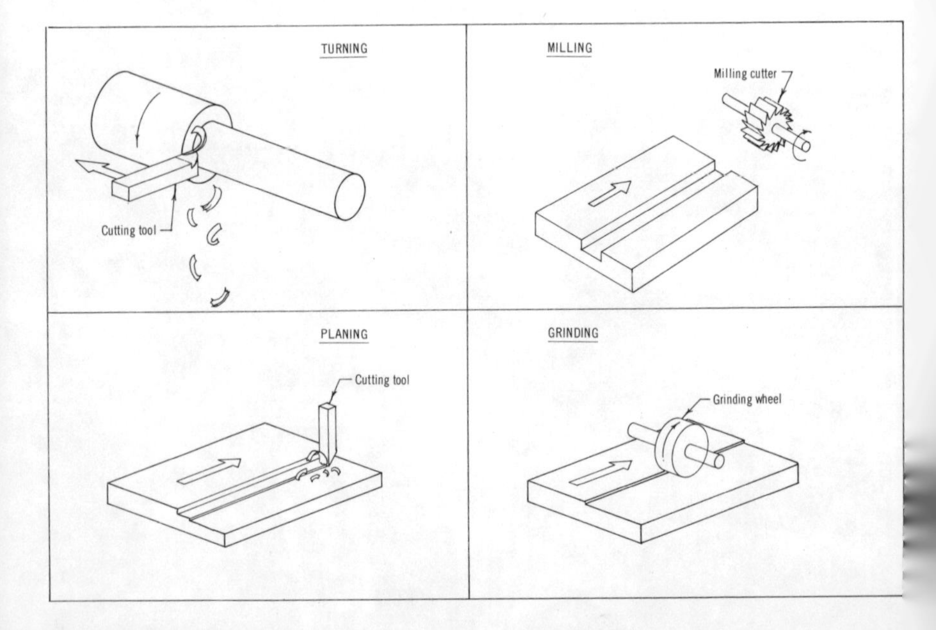

Fig. 8-10. **Nearly all of the main processes for working metal have been used to manufacture this assortment of items. (*Anaconda Brass Division*)**

few thousandths of an inch. Also, instead of using only wrought iron and simple steel, he would have a choice of the thousands of alloys developed in the past two hundred years—machine steels, strong alloy steels, heat-treated steels, special low-friction bearing metals, brass, bronze, aluminum, and many others.

Some of the metals and alloys available to us today, along with some of the reasons for choosing one metal over another, are the topics of the next chapter.

CHAPTER 9

WHAT KIND OF METAL?

More than thirty years ago, a scientist on the radio serial "Buck Rogers in the 25th Century" invented a very special metal. He built spaceships, rocket engines, and just about everything else from it. This fantastic metal weighed practically nothing, was so strong that it could not be broken, could not be melted by the hottest rocket exhaust, and was impervious to the most powerful bullet or projectile. The imaginary scientist of the twenty-fifth century called his imaginary metal "impervium."

Do we have—or are we likely to develop—a super metal like "impervium" that is so superior in every way that we can use it for everything?

The answer seems to be that no one metal or alloy could ever serve all of the thousands of uses we have found for metal. We now have some two dozen elements that are commercially-useful metals. Thousands of alloys of these metals have been developed, and more are being found as new needs arise. Each of them will do certain jobs better than any other; none of them can do everything. Metals (and alloys) are all different in their weight, ductility, strength, hardness, ability to conduct electricity, ability to resist corrosion, ability to withstand high (or low) temperatures, and so on.

Almost any machine needs parts that are made of many different metals and alloys, and which do different jobs. For instance, a bulldozer needs soft metal for bearings that must run in a slippery bath of oil, and hard, tough steel for wheels and tracks that can grind along in sand and dirt without harm. It needs fine steel that can be easily machined into polished, close-fitting engine parts, and rough, strong metal for the blade that will not wear away as it tears into rock and soil. For its electrical system, the bulldozer needs metals that will conduct electricity well, and for its cooling system, metals that are good conductors of heat.

A metal selected to do a particular job must be available in sufficient quantity and at an acceptable price. Some metals are plentiful and cheap, while others are scarcer and more expensive. Some rare or "precious" metals are very costly.

For some purposes, the most expensive metal might be just the right one to use. Electrical contacts and switches aboard a spacecraft might be made best of platinum or silver, for example. Only a tiny button of the precious metal is needed here and there and only a few ounces or a few pounds might be needed for the whole spacecraft. Enough of these metals is available for such purposes, while the high cost is warranted by the value placed on the space program.

But for building a large structure, such as a bridge, thousands of tons of metal might be needed. We could build the bridge of bronze or of gleaming stainless steel. It would be quite strong enough, beautiful to look at, and would never need painting because it would not rust. A bridge of these metals, however, would cost many times as much as a bridge made of more ordinary metal. Also, the metals in these alloys are not plentiful enough to use in such huge quantities. They are more urgently needed for other purposes. So we build bridges of an ordinary steel that is plentiful, relatively inexpensive, and yet strong enough for the job. Because such steel will rust, the bridge will need more painting and other maintenance over the years and may have a shorter life than a bridge of bronze or stainless steel.

Sometimes a new metal that becomes more plentiful and cheaper will replace one that is becoming scarcer and more costly. Copper has been used for over a hundred years for electrical wires of all types. Copper was chosen for this job because it is an excellent conductor of electricity. The only better metal for conducting electricity is silver, which is too soft, scarce, and expensive to be used for such purposes. Copper is heavy, though, and the large cables used to carry high-voltage power across the country are very expensive. Aluminum is a good conductor of electricity, though not as good as either silver or copper. But aluminum weighs less than ⅓ as much as copper, so that larger cables can be used to make up for the loss in conductivity. Because aluminum is lighter and cheaper and can carry electricity almost as well as copper, aluminum cables instead of copper cables are being strung between high-tension towers all across the country. Aluminum is slowly replacing copper for many other electrical uses, such as house wiring and electric motors.

The different weights of different metals often is a factor affecting selection. For a bulldozer, big heavy chunks of steel can be used for strong parts because bulldozers must be heavy to do their job. An airplane, on the other hand, must be light in weight to do its job. Metal parts must be carefully designed for lightness. Lightweight metals, such as aluminum, magnesium, and titanium, are used as often as possible. Lightweight metals often cost more than heavier metals. As a result, an airplane may cost ten times as much, per pound, as an automobile.

For some purposes, a metal that will not corrode or rust must be used. The pipes, kettles, and tanks used by the food industry for pasteurizing milk or in which to cook soup must be made of metals that will not affect—or be affected by—the food passing through them. The metal, too, must have a bright, durable surface that can be easily cleaned and sterilized between uses. A good grade of stainless steel is expensive, but is essential for uses such as this. There are many different kinds of stainless steels, containing various amounts of chromium, nickel, and manganese.

For some uses, a metal is needed that expands or contracts very little with changes in temperatures. Such metals are used in watches and other precision instruments to keep their accuracy from being affected by temperature. An ordinary metal surveyor's measuring tape would change its length so much between a hot and cold day that a surveyor would need to correct his measurements for this temperature change. A measuring tape made of "Invar," an alloy of nickel and iron, would change its length so little that no correction would ordinarily be needed.

Have you ever seen a spoon melt in a cup of hot coffee? If your spoon was made of an alloy known as *Wood's metal*, it would do just that. The spoon would melt down into a blob of metal in the bottom of the cup. This important metal is used to make heat-sensitive valves for automatic sprinkler systems. When a fire in a building raises the temperature at the valve to about 160°F (71°C), this metal melts and lets the water spray out into the room to put out the fire. Wood's metal is an alloy of bismuth, cadmium, lead, and tin, in the proportions that will give the lowest possible melting point for the combination—far lower than the melting point of any of the elements by itself (Fig. 9-1).

Such an alloy is called a *eutectic*. Most soldering and brazing alloys are eutectic alloys designed to melt at low temperatures in order to make the soldering or brazing job easier.

Ordinary iron and steel are the metals most often thought of as being magnetic. But special alloys have been developed that make much more powerful, permanent magnets. One such alloy is "Alnico," which contains 8 percent aluminum, 14 percent nickel, 24 percent cobalt, 3 percent copper, and 51 percent iron. These magnets are widely used in electric motors and other electrical devices.

Rocket engines and jet engines require some very special kinds of alloys that will keep their strength at very high tem-

***Fig. 9-1.* The alloy "Wood's metal" has a melting temperature far lower than any of its constituents.**

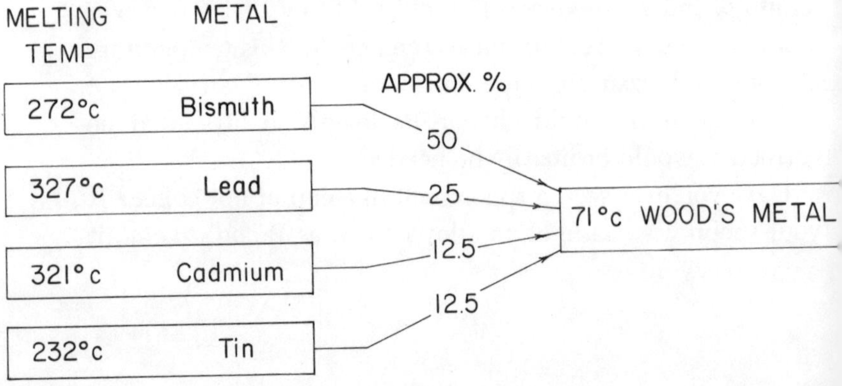

***Fig.* 9-2. Only very special alloys that retain their strength at high temperatures can be used in rocket engines. (*NASA*)**

peratures (Fig. 9-2). The higher the temperature that can be used in an engine, the greater the amount of power that can be squeezed from a given amount of fuel. Unfortunately, most ordinary metals tend to become soft and weak and to oxidize rapidly when heated very hot. Nickel and chromium have been found to be especially resistant to high temperatures. Special alloys of these metals have been developed for turbine blades that operate up to nearly 1,900°F (1,037°C). Above this red-hot temperature, some sort of cooling is needed, or some new material such as carbon fiber, or a combination of metal and ceramics will have to be used.

Corrosion is often one of the most important considerations in choosing a metal. Iron rust—that reddish, powdery scale that forms on iron or steel—is familiar. Probably everyone has at one time or another left a tool or toy out in the rain and found that it has been nearly ruined by rust in only a few days. Rusting is one form of corrosion.

The corrosion of metals costs hundreds of millions of dollars every year, and the cost would be even higher if more millions were not spent in preventing corrosion.

We learned in earlier chapters that most metallic elements exist naturally in the earth in the oxidized condition, having been put in that condition through contact with the earth's air and water over millions of years. We might expect that after we have extracted metal from its ore, it would have a strong tendency to go back to its natural oxidized state as quickly as possible.

Fortunately, some metals, when they are attacked by oxygen, form tight, durable coatings of metal oxide that protect the surface from further oxidation (this process is sometimes called dry corrosion, or *tarnishing*). The surface of pure aluminum, for example, oxidizes back to *alumina* (Al_2O_3), a dull gray film. Alumina is chemically similar to aluminum ore. It is a stable compound that is hard and durable. Because the film of alumina is stuck tightly to the surface, it is excellent protection for the aluminum metal beneath.

Even iron, if it is kept dry, will form a coating of iron oxide that will protect its surface. If the earth's atmosphere were as dry as the atmosphere of Mars, iron would corrode very, very slowly. Some of the desert areas of the United States are so dry that iron rusts very slowly there. Moisture is the main villian that attacks iron—with the help of foreign particles brought by air pollution. Iron oxide and many kinds of pollution particles readily absorb moisture from the air. Only a thin film of moisture is needed to begin the commonest and most destructive kind of corrosion called *electrolytic corrosion.*

The chemical cell is an example of practical use of electrolytic corrosion. Two metal electrodes are suspended in a weak acid as shown in Fig. 9-3. The acid attacks the zinc electrode and dissolves away zinc atoms. When the zinc atoms enter the water, each becomes an ion (Zn^{++}) by leaving behind two electrons. The zinc electrode thus becomes negatively charged. The acid does not dissolve many copper atoms; therefore, the copper plate has only a very small negative charge. (Copper, you will recall from Fig. 2-2 is lower in the chemical-activity series than zinc; hence, it is the zinc that is dissolved into the solution.) If a wire is connected between the two electrodes, the crowded electrons on the zinc plate flow along the wire to the copper plate as shown in the figure. These electrons are picked off of the plate by hydrogen ions (H^+) from the acid. The ions of hydrogen that acquire electrons become neutral atoms of hydrogen and bubble to the surface as molecules of hydrogen gas. As the electrons leave the zinc plate through the wire, more zinc atoms can dissolve, leaving more electrons behind. In this way the electric current will continue to flow until all the zinc is dissolved.

***Fig.* 9-3. The chemical cell is an example of the practical use of electrolytic corrosion.**

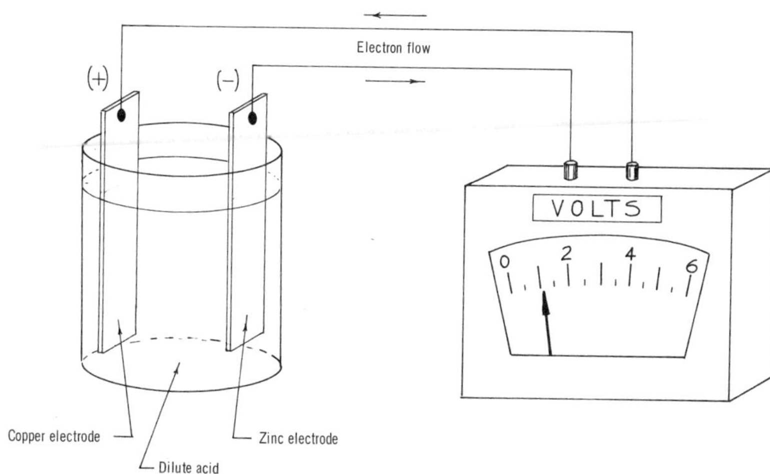

Two dissimilar metals that are in contact (or in close proximity) in the presence of a thin film of moisture will form a similar electrolytic cell in which a weak electric current will be produced. As the electric current flows between the two metals, the more active one will be eaten away. Even a single piece of metal can be corroded in this way in salt water. Impurities in the metal and even differences in the metal between the crystals and the crystal boundaries can set up tiny electrolytic cells. When electric current flows between any two points, even though it is an extremely weak current, metal is corroded away.

Electrolytic corrosion is especially destructive on boats in sea water. Because sea water is a good electrolyte, a relatively strong current of electricity will flow between an iron rudder and a bronze propeller, with the iron rudder being quickly eaten away. For this reason, only one kind of metal should be used in the hull of a boat whenever possible. If the propeller, rudder, all fastenings, etc., are made of bronze, electrolytic corrosion will be held to a minimum.

There are a number of ways to avoid metal corrosion. One is to use metals that are almost unaffected by the chemicals which may attack them. Silver and gold make good fillings and crowns for teeth, because these metals will not be corroded by the continuous immersion in the digestive fluids in the mouth.

Another method of protecting metal is to allow (or help) the surface to become coated with a durable oxide coating. In a process called *anodizing*, the surface of aluminum can be purposely oxidized to form a hard coating of aluminum oxide that will afford it protection from further corrosion under most circumstances.

A metal may be plated or "clad" with another metal that is more resistant to corrosion. A thin sheet of corrosion-resistant aluminum is sometimes rolled (and bonded) onto a sheet of steel for this purpose. The aluminum develops an oxide coating on its outer surface that protects it and the steel beneath from further corrosion. Some alloys of aluminum are very strong, but have poor resistance to corrosion. A thin sheet of pure aluminum is sometimes bonded to the alloy to protect it from corrosion.

Chromium is a very durable metal, but when plated on steel, it is porous and allows corrosion agents to eventually penetrate to the steel beneath. In much of the "chrome plating" we see on autmobiles, the steel is first electroplated with a more protective metal, such as copper or nickel (or both). Then the chromium is plated on top to provide a hard, attractive coating that will last a long time.

Perhaps the strangest method of protecting metal is to add to it, in an alloy, a metal that will protect it from the *inside*. If chromium is added to steel in an amount of about 18 percent, the chromium atoms in the metal can oxidize at the surface and form a tough protective film over the steel. This is the secret of the alloys called "stainless steels."

Sometimes one metal can be used as a "sacrificial" metal to protect another. As pointed out above, if bronze and iron are involved in an electrolytic reaction on a boat hull, the iron will be quickly attacked and corroded away. Iron is chemically more active than bronze. If a metal higher than iron on the chemical activity table in Fig. 2-2 (such as zinc) is available, that metal will be attacked instead of the iron. On boat hulls, then, where the use of iron and bronze cannot be avoided, a chunk of bare zinc attached to the iron will "sacrifice" itself to electrolytic attack, thereby protecting the iron. When the zinc is corroded away, it can easily be replaced.

A very familiar method of protecting metal from corrosion

is painting. Although special paints can usually give considerable protection, paint tends to be porous and will eventually let enough moisture through to the metal to start corrosion beneath the paint. This is why structures in the weather, such as steel bridges, must be scraped to remove the old paint and then repainted every few years. Some plastic coatings have been developed that give longer protection.

The chemical reactions that cause corrosion are not always harmful but are often useful for many purposes. The lead-acid storage battery used in automobiles, electroplating a silver teapot, cleaning metal, cutting steel with an oxygen torch, and the photographer's flashbulb are all practical uses of corrosion reactions. None of these would be possible if metals did not react vigorously with oxygen and other elements. But the trick is to prevent these reactions from happening where we don't want them. To protect against corrosion, we spend hundreds of millions of dollars each year to keep metal from crumbling and dissolving back into the useless minerals from which it was smelted in the first place.

When we look at all the different things that metals are asked to do, we can doubt that even Buck Roger's twenty-fifth-century metal "impervium" could fill all the requirements. Engineers today still talk of such a metal, however, when they are designing a new gadget and need a metal with qualities far beyond those of any available material. We can guess that they are not very serious, though, when they say they will build their gadget out of that modern-day wonder metal called *Unobtainium*.

CHAPTER 10

ALCHEMISTS AND ATOMS

One metal that has been important throughout history, not because it is plentiful, but because it is scarce, is gold. Because of its beauty and desirability, gold has been used for jewelry and works of art. Because of its rarity and great value, gold has been important as money—a medium of exchange in world commerce. The quest for this beautiful and valuable metal has led people to explore new oceans and settle distant lands. The quest for gold has also led men into less constructive activity, such as robbery, piracy, war, and murder.

Perhaps the strangest method of seeking gold—and the most harmless—was the attempt to make it in the laboratory from a common metal, such as lead. This idea probably began with the ancient Greeks and gripped the imagination of the world for nearly 2,000 years. When we examine early ideas about the nature of matter we can see how people came to believe that lead might be changed into gold.

The ancient Greeks were the first people to record their ideas about the world. The very earliest of the Greek thinkers believed that the world was made up of only one basic substance, and that different kinds of matter were simply this same substance in different forms. One early Greek philosopher thought that this basic substance was water; later, another believed it to be fire. Still later, another philospher declared that matter was made up of not one but *four* basic particles: earth, air, fire, and water. Although this view of matter may sound ridiculous to us today, if we examine it we will see that this view fits some everyday observations very well. When we put metal ore (earth) together with fire, for example, we get metal. Therefore, metal must be a combination of fire and earth. A tree uses soil (earth), water, and air to grow when the light of the sun (fire) falls upon it. Therefore, wood must be a combination of earth, water, air, and fire. Want further proof? When wood is burned, what is released? Fire (heat), earth (ashes), air (smoke), and water (vapor in the smoke).

The great Greek scientist, Aristotle, accepted the idea of four basic elements and decided that there were also four fundamental qualities that give a substance its form: heat, cold, moisture, and dryness. He also believed that these qualities could combine to make physical substances: heat and dryness could form fire, heat and moisture could produce air, and so on. The stature of Aristotle was so great that his ideas were believed for many centuries.

If the various kinds of matter that can be seen all around us are the result of various combinations of these basic elements, then it should be possible to find a process by which *any* substance—including gold—could be made. This was the reasoning that set the philosophers, scientists, and even uneducated people, to the task of trying to convert base metal into gold. The alchemists, as they were called, kept at the task of grinding, mixing, boiling, and melting all sorts of strange concoctions in their laboratories, in spite of continual failure, for many centuries (Fig. 10-1).

It was during the Middle Ages that alchemy reached its peak. All over Europe, kings and princes who needed gold to replenish the treasuries of their kingdoms eagerly sponsored the search for the process for transmuting lead into gold. So great were the rewards offered for success that some alchemists rigged phony demonstrations in which they *seemed* to change base metal into gold. They were all exposed in time, if not immediately. False claims became so numerous that exasperated rulers began to deal harshly with

Fig. 10-1. **Medieval alchemist's laboratory. (*New York Public Library Picture Collection*)**

those who could not produce what they had claimed. A number of frauds were hanged or burned at the stake as a warning to others. Still the search went on. No one seemed worried that an easy process for changing base metal into gold would make gold so plentiful that it would no longer be especially valuable.

Some of the foremost scientists of later days believed that transmutation of one element into another was possible. Sir Isaac Newton (1642–1727), one of the greatest scientists of all time, tried his own hand at transmutation experiments around 1700. He was no more successful than the rest. In the latter part of the nineteenth century, less than 100 years ago, Charles M. Hall, whom we met in Chapter 3 as the inventor of the first practical process for smelting aluminum, also dabbled in transmutation experiments.

The time and effort spent through the centuries by the alchemists on the problem of transmutation fortunately was far from a total loss. Alchemists learned how to perform such experimental operations as measuring, weighing, and distilling, as well as how to build and operate laboratory equipment. They developed methods of investigation, and slowly the science of chemistry began to emerge from the smelly pots and crucibles of the alchemist. Discoveries proceeded rapidly. At about the time of Columbus, chemists had classified acids, bases, and salts and had identified alcohol and many other substances. A good number of metals were known by this time, and many of them were in commercial use.

But no one yet knew that the atom was like, what air consisted of, or what chemical elements they were dealing with in their work. No one yet knew what chemical process took place when a metal was smelted from an ore. Around

1670, J. J. Becher (1635–1682), a German chemist, proposed a new theory which assumed the existence of something called *phlogiston,* a sort of "fire principle." A metal locked up in an ore, this theory said, must have phlogiston added to it in order to be released as a free metal. Conversely, as a metal rusted away into an oxide, it did so because it was losing its phlogiston. This theory, of course, was exactly backwards from the smelting and rusting processes as they actually take place with oxygen (iron ore *releases* oxygen to yield iron; iron *gains* oxygen as it rusts). Though we may smile at the phlogiston theory today, it has been called the first great generalization in chemistry. Science was beginning to probe the secrets of chemistry in earnest.

Early in the 1800s, John Dalton (1766–1844) proposed the first truly chemical atomic theory. He suggested that every element had atoms of a particular size and weight. Experiments around this time began to support his atomic theory and to show also that the forces holding elements together in compounds are electrical in nature.

During this period chemists began to classify the elements that were then known. Atomic weights were determined for most elements, and valence (the number of single bonds that an element can form when it combines in compounds) was understood. The idea that there was a certain order among the elements began to grow. As the characteristics of the known elements were examined, in order of increasing atomic weight, for example, investigators found that certain characteristics were repeated in a fairly regular order. This discovery led several people to attempt to arrange the elements in a chart or table in a way that would show family relationships. In 1869, the Russian chemist, Dmitri Mendeleev (1834–1907) invented the *periodic table.*

Mendeleev later revised it, and new revisions have been made by others as more was learned and more elements were discovered.

A modern form of the periodic table appears as Fig. 10-2. In this table the elements are listed in horizontal rows, from left to right, in order of their atomic number. (Early tables used atomic weights. The atomic number, which is the number of protons in the atomic nucleus, came into use some decades later and gives almost the same order.) The rows are then stacked, one on top of the other, in a way that places elements having similar chemical and physical characteristics in vertical "families." An element in any vertical column is usually more like others in that vertical column than it is like others in the table. For example, the metals copper, silver and gold, in family IB, are more like each other than they are like any of the metals on either side of that column. Elements in any family will usually form compounds like those formed by the others in that same family. The metals lithium (Li), sodium (Na), and potassium (K) in family 1A are very active metals that are very much alike and also form chemical compounds that are similar.

The elements to the right of the heavy line that jogs down through the right half of the table are nonmetals while the elements to the left of this line, with the exception of hydrogen, are considered metals. The distinction between metals and nonmetals, however, is not always completely clear. For example, the element silicon, which is considered to be a nonmetal, can be polished to a metallic luster and has a greater electrical conductivity than the element below it in the table, germanium, which is considered to be a metal. The metal bismuth will shatter like glass when struck with a hammer, while pure tin loses its metallic structure to become

Fig. 10-2. Periodic Table of the Elements. (*Copyright 1964. E. H. Sargent & Co.*)

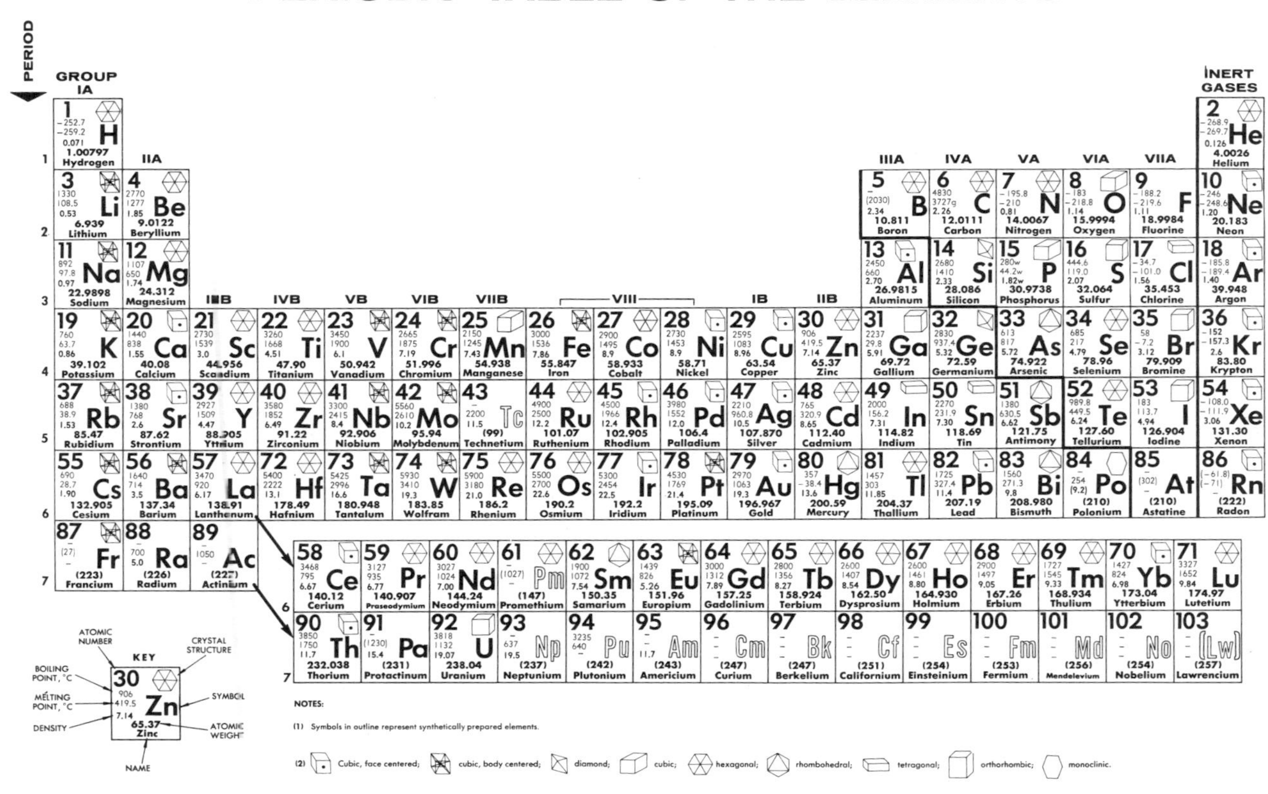

a powder at low temperatures. Because they have some of the characteristics of metals, elements such as carbon and silicon are sometimes called *metalloids*.

At the time Mendeleev fitted the elements into his periodic table, no one knew what made certain elements behave in their characteristic ways. Several decades later, toward the end of the nineteenth century, discoveries were made that led to a new understanding of the atom. These discoveries explained chemical activity and showed why and how the periodic table worked. Both discoveries were concerned with metal and with the (then) new science, dealing with electricity.

The first discovery began in 1883 with an experiment by the American inventor, Thomas Edison (1847–1931). He placed an extra wire in the evacuated glass envelopes of one of his electric lamps. He found that when this wire was connected to the positive side of the electrical circuit, a current flowed through the vacuum between the wire and the hot lamp filament. When the wire was connected to the negative side of the circuit, to reverse the current, no current flowed. Everyone knew that an electric current would flow through a wire, but an electric current flowing through a *vacuum* was something new. And why did it flow in only one direction?

Edison did not understand what he had discovered nor did he see any practical use for it, so he simply recorded the phenomenon and set it aside. A few years later, a British scientist, Sir Joseph Thomson (1856–1940), completed the discovery begun by Edison. He investigated the visible "rays" that had been observed to occur in an evacuated glass tube when an electrical voltage was connected across it. He concluded that the "ray" was actually a flow of tiny particles that are contained in all atoms. He called these particles "corpuscles"; they are now called electrons. He found later that these same particles could be produced in a tube by heating a metal in a vacuum.

We know now that when the atoms of a metal become very hot, the speeds of the electrons orbiting the atom become so high that some of them leave the outer electron belts and fly off into space to form an electron cloud around the hot metal. The mysterious electric current that Edison had found flowing through the vacuum in his electric lamp was, of course, a flow of these loose electrons from the cloud around the hot filament to the positively-charged wire.

With his discovery of the electron, Thomson not only explained the Edison effect but also smashed the widely held idea that atoms were the smallest particles in the universe. The atom could not be the smallest particle, since he had shown that it contains still smaller particles, the electrons. This discovery opened up the new field of atomic physics, better known as *nuclear physics,* which deals mainly with atomic particles and with what goes on inside the atom.

Other very important events of that period, which we shall mention only in passing, were the discovery of radioactivity by Antoine Henri Becquerel (1852–1908) and the work on radioactive elements by Pierre and Marie Curie. This work further advanced the understanding of the atom and showed that radioactive elements spontaneously eject atomic particles and energy and change themselves slowly into other elements. This discovery completely overturned the accepted view that the atoms of an element were unchanging and unchangeable.

A British scientist, Ernest Rutherford (1871–1937), followed the work of Thomson to probe further the mysteries of the flow of electricity through gases and the new phenomenon of radioactivity. He devised a new experiment to find out what atoms are like by bombarding them with fast-moving atomic particles ejected from a radioactive gas. He used a very thin gold foil as his target (because gold is so malleable it could be hammered into an extremely thin sheet). Gold atoms, like all other metal atoms, are tightly stacked in crystals with very little space between. Even in this extremely thin foil, the atoms might be expected to form a solid wall perhaps 100 atoms thick.

When Rutherford bombarded this sheet of gold, most of his atomic bullets went right through the foil without being slowed or deflected at all—almost as though the foil was not even there. But a few of the particles were deflected as they passed through the foil; that is, they acted as though they had struck something a glancing blow. A very small number of other particles bounced back from the foil as though they had struck something very heavy and very hard (Fig. 10-3).

This latter result especially astonished Rutherford. Because his atomic bullets were moving at a speed of 10,000 miles per second, turning them back with gold foil was, he said, like stopping an artillery shell with a piece of tissue paper. He concluded that an atom must consist of a small central body, or nucleus, with a positive electrical charge, since it repelled his positively-charged bullets. This nucleus must be very heavy, since the bombarding particles were bounced back so readily. But most important, the nucleus must be extremely small, since it was struck so seldom by a bombarding particle, even though the foil was made up of 100 layers of atoms or more. He concluded that, except for this tiny nucleus, the atom must be mostly empty space in

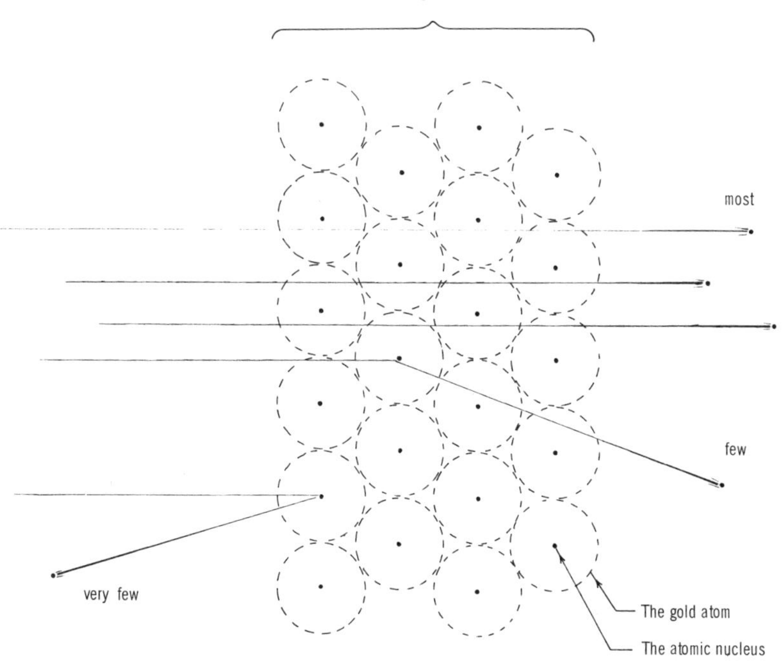

Fig. 10-3. **In Rutherford's gold-foil experiment, most of the bombarding particles passed straight through the foil, while a few were deflected and a very few turned back.**

which the electrons move. We now know from later measurements that the nucleus is only about 1/100,000 as large as the atom. This means that if an atomic nucleus were increased to the size of the period at the end of this sentence, the outer edge of the atom, where the outer electrons orbit, would be about 250 feet away. Except for the electrons, which are even smaller than the nucleus, the rest of the atom is empty space.

Out of these and other discoveries during the last half of the nineteenth century and the early years of the twentieth century came a new idea of the atom, put into a new "model" by Niels Bohr (1885–1962) in 1913. The atom, as he saw it, consisted of a heavy central nucleus and orbiting electrons arranged in shells. Further development of this view of the atom showed that the chemical characteristics of an element is related to the number of electrons in its outer shell. The reason for the similarities between members of families of elements in the periodic table at last became clear. In general, members of these families have the same number of electrons in their outer electron shells. An understanding of the role of electrons in chemical bonds, crystal structure, and the electrical conductivity that has already been mentioned in earlier chapters came out of this work.

Perhaps the most startling discovery to come out of radioactivity and the new view of the atom was the discovery that the ancient alchemists were right after all—one element *can* be changed into another! The alchemist was in error only in believing that it could be done easily and in the crude, smelly potions brewed in his laboratory.

The principal element of all stars is hydrogen. We now know that the sun obtains its energy from the transmutation of hydrogen into helium. Six-hundred fifty million tons of hydrogen are transmuted (or "fused") each second into the heavier element helium, with the energy from the reaction pouring outward into space as heat and light. This is the process that we hope to harness in fusion power plants here on earth to generate electrical power.

The heavy elements in the Universe are the result of fusion of lighter elements into heavier ones in the unbelievably high temperatures and pressures in the centers of stars. Because our solar system contains a relatively large amount of heavy elements, scientists believe that the materials of which our sun and its planets are made were part of another star many billions of years ago. When that star burned out and exploded, as stars sometimes do when they grow old, it formed a cloud of dust and gas that later came together again to form our sun and its solar system. Actually, the amount of heavy elements found in our solar system is large enough to suggest that this material may have gone through the atomic furnaces of *several* stars before finally forming our solar system. In each star the amount of heavy elements increased. Each time the star grew old and exploded, a new star was formed from the debris, until finally our sun was formed.

The uranium metal that we used in nuclear power plants was built up from simpler elements in this way. When we cause the uranium to split, or fission, into simpler elements, we release energy that came from the atomic furnaces of stars that existed many billions of years before our sun was formed.

Unfortunately for the gold-hungry kings of the Middle Ages, changing another element into gold can be done only under very special conditions, such as the crushing pressures and searing temperatures found in the center of a star. These conditions, plus the large amount of energy required to achieve a transmutation, make it unlikely that it will ever be practical to manufacture gold for a king's treasury or for any other purpose.

CHAPTER 11

ELECTRONS, METALS, AND ELECTRONICS

The electron is not just the lightest stable elementary particle known to exist in matter. It also appears to be the busiest particle, and one that seems able to do almost anything. The electron is sometimes a particle, sometimes a wave. It is usually tied to an atom by the electrical attraction of the nucleus, but sometimes exists alone in space. It sometimes seems to move by drifting slowly, but often tears along at nearly the speed of light. The electron is extremely small in size, yet zips around an atomic nucleus to form a moving shield that makes the atom 100,000 times larger than the electron itself. The electron gives the atom its particular chemical characteristics and is largely responsible for metal being the strong, lustrous, flexible, ductile material that it is.

For the first 7,000 years that metal was used on earth, its electrons were not asked to do more than their normal job of making metal what it is. About 200 years ago, scientists began to use the electrons in metals in new ways. Sometimes in metal, sometimes off on their own, electrons became the carriers of a new form of energy called electricity. The benefits that electricity and electronics have brought us, especially in the past fifty years, hardly need to be described.

The electrical age began with the invention of the first electric battery in 1800. In honor of its inventor, Alexander Volta, the battery is often called a *voltaic pile.* It made available the first practical amounts of current electricity. Until this time, only the static generator and the Leyden jar had been available to generate and store small amounts of static electricity for brief experiments. Volta's battery was made of disks of copper and zinc separated by damp material such as cloth soaked in a salt or acid solution. Following Volta's invention, batteries using various metals and solutions were used as the principal sources of electricity for nearly a hundred years—until the electric generator was developed. They are still in use today as portable sources of electrical energy, such as the automobile battery and the flashlight cell. The source of the electrical energy produced by a battery is not the metal itself, but the chemical reaction between the metal and the acid.

With larger quantities of electricity available from the chemical battery, scientists plunged ahead with new experiments in electricity. New discoveries and new inventions followed each other in rapid succession. The experiments of Thomas Edison and Lord Thomson, described in the previous chapter, were among the most important. They showed that the electron, one of the particles that makes up the atom, was the carrier of electricity through conductors, and, under some conditions, through empty space. They also showed that a heated metal filament will be surrounded by a cloud of electrons.

An English inventor, Sir John Fleming (1849–1945), used these results to invent the two-electrode vacuum tube known as a *diode*. This device, described in detail in Fig. 11-1, was a one-way valve that used the cloud of electrons to transmit electricity through the tube in only one direction. The main use of this simple tube is to change an alternating current into a direct current.

An American engineer, Lee De Forest (1873–1961), improved on Fleming's idea by adding a third electrode, in the form of a metal grid, between the other two. This vacuum tube, called a triode, also serves as an electrical valve. When a weak signal is used to create an electric charge on the grid, this charge controls the flow of electrons between the other electrodes in the tube. (See Fig. 11-2 for details.) As a result, the electrical signal emerging from the vacuum tube can be made not only a copy of the original weak signal, but very

***Fig. 11-1.* The cloud of electrons thrown off by a heated metal filament makes possible the electronic tube. This two-element tube, or diode, will pass an electric current in only one direction. (Electrical circuits are not shown on the diagram.)**

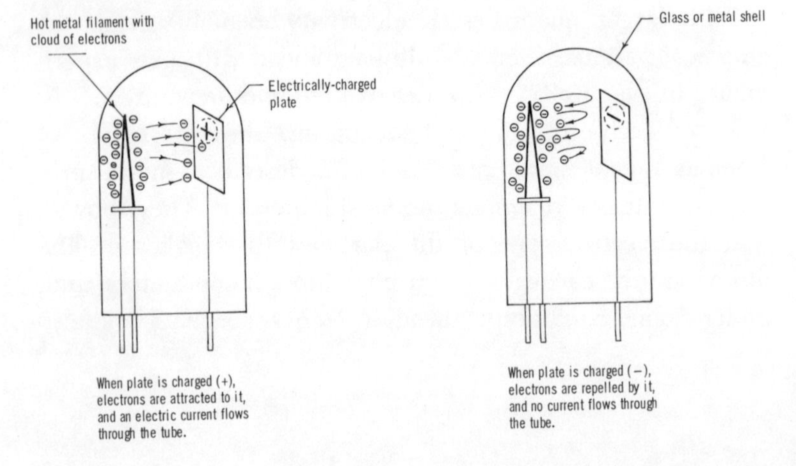

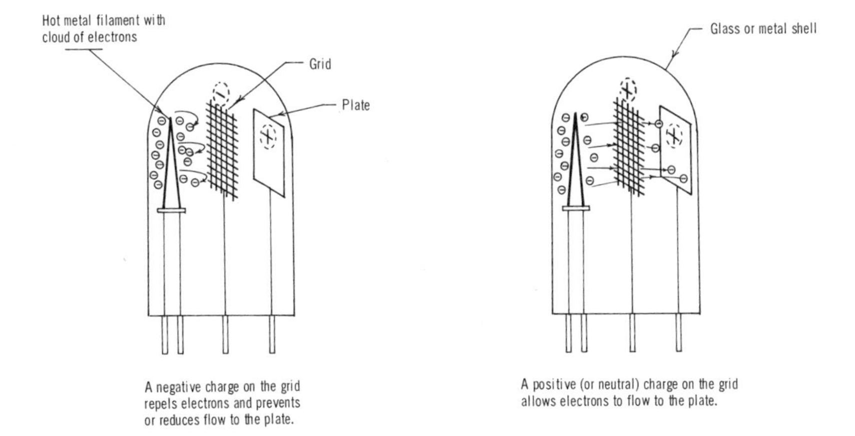

Fig. 11-2. **The three-element electronic tube, or triode, contains a charged grid that acts as valve to control the flow of electrons between filament and plate. A weak electrical signal placed on the grid can be amplified (made stronger) by this tube. (Electrical circuits are not shown on the diagram.)**

much stronger as well. This ability to "amplify" an electrical signal is essential for radio, long-distance telephone, and other electronic devices.

These inventions, and the whole technology of electronics, were made possible by the ability of heated metal to boil its loose electrons out into the space around it. There is evidence that the electrons in metal continually plunge through the surface of cold metal from the underside, flying out beyond the last row of atoms for a short distance before being pulled back by the attraction of these atoms. When metal is heated, the speed of the electrons increase. When the speed becomes great enough, the electron can escape from the forces that hold it and fly off into space on its own.

As we might expect, other forms of energy besides heat can also break electrons loose from metal. Only a few years after Edison's discovery, other scientists found that radiant energy—such as ultraviolet light or visible light—falling on cold metal can excite electrons enough to allow them to escape. The photoelectric tube, (Fig. 11-3) based upon this effect, is used in photographic light meters, television cameras, and many other simpler devices such as fire (smoke) alarms, door openers, and traffic-counting equipment.

Because vacuum tubes are heavy, fragile, and use a great deal of power to heat their filaments red-hot, electronics engineers longed for smaller, more efficient devices that would do the job. They found what they needed in certain metals and near-metals called *semiconductors*. Two important semiconductors are the metal germanium and the nonmetal silicon. Both, you will note, are in the same vertical family in the periodic table (Fig. 10-2), which also includes carbon, tin, and lead. Both are poor conductors of electricity, and poor insulators as well. (Surprisingly, the nonmetal silicon is a better conductor than the metal germanium!)

***Fig. 11-3.* The photoelectric cell.**

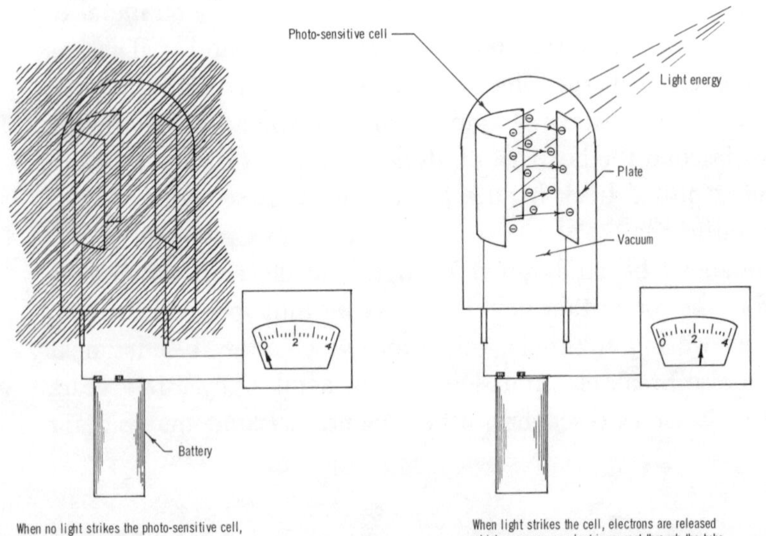

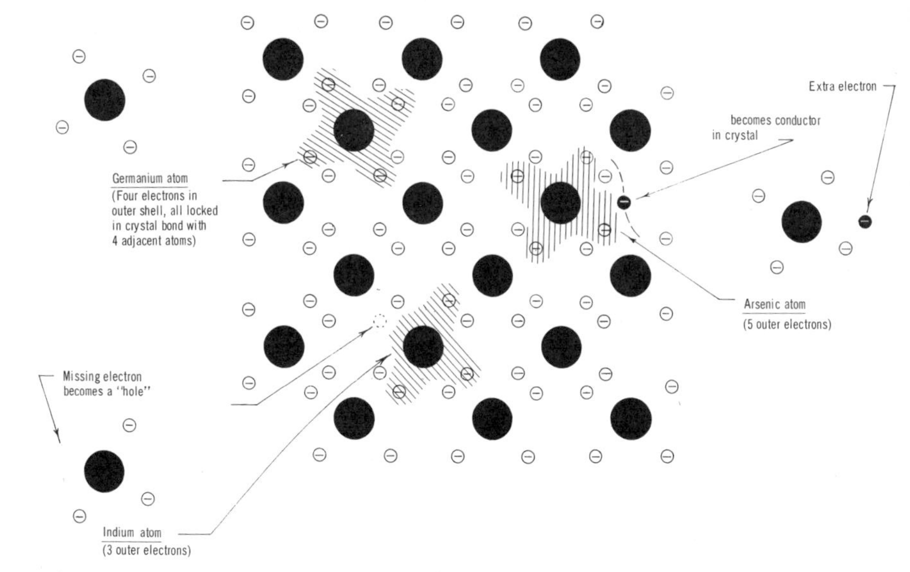

Fig. 11-4. **A crystal of germanium atoms with two "impurity" atoms added. Only outer-shell electrons are shown.**

Semiconductors are poor conductors of electricity because they have few or no free electrons. Because the germanium atom has four electrons in its outer (incomplete) electron shell, we might expect it to be a conductor. But germanium atoms arrange themselves in crystals in a way that is different from most metals. None of the electrons are really free to move through the solid, but instead are tied up in atom-to-atom bonds in the crystal structure.

But when an atom of another element that has one more free electron in its outer shell—such as antimony or arsenic—is added to the germanium, this element will bring an *extra* electron into the crystal which is not needed in the structure and which will be left free to roam within the crystals. Such electrons make the germanium a fair conductor of electricity (Fig. 11-4). This semiconductor is called a "negative carrier," or *n-type material.* If a different atom which has *fewer* free electrons in its outer shell—such as indium—is added to the germanium instead, it brings a shortage of an electron into the crystal structure which leaves an empty space in the crystal called a *hole.* Holes make the germanium semiconductor act as an absorber of electrons. This semiconductor is called a "positive carrier," or *p-type material.* Adding impurities to semiconductors for these purposes is called *doping;* normally an extremely small amount of the impurity element—perhaps one part per million—is added.

When these two kinds of semiconductors are placed together, they make a unit called a *diode.* Whether current will flow through the diode or not depends upon whether electrons are available at the junction of the two materials. As Fig. 11-5 shows, these electrons are available only when the electrical voltage is applied in one direction. Like the vacuum-tube diode, then, the semiconductor diode will con-

Fig. 11-5. **The semiconductor diode permits an electric current to move through it in only one direction.**

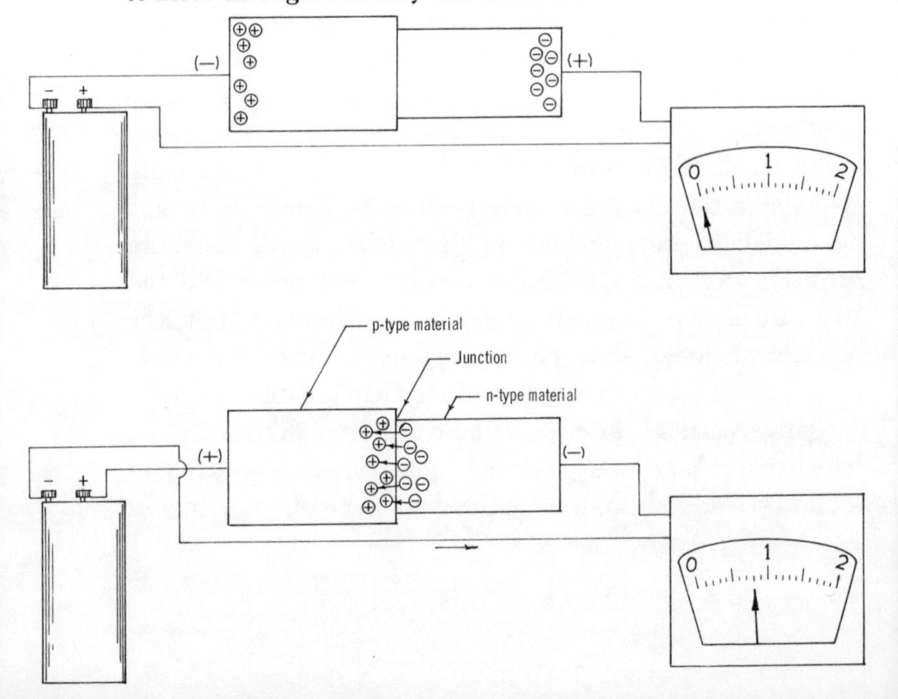

duct an electric current in only one direction and will therefore change an alternating current into a direct current.

Semiconductors can be arranged in ways that will allow a weak electrical signal to be amplified into a large one. In these devices, called *transistors,* the current can flow only when the weak signal provides the electrons to cross the junction. Transistors can be used nearly everywhere in electronic circuits instead of vacuum tubes. The tiny size of semiconductor devices has made it possible to reduce electronic equipment that would formerly have occupied a room into a package that can be carried in one hand. Because no red-hot filament is needed, transistors use much less electric power and create much less heat than do vacuum tubes. Transistors have made possible pocket radios, pocket calculators, electronic wrist watches, and many other tiny electronic devices.

Having learned that both heat and radiant energy can break electrons loose from ordinary metal, we might suspect that heat and radiant energy might also have an affect upon the availability of electrons in semiconductors. This is exactly the case, and many very useful devices are built that utilize these effects. For example, radiant energy (light) falling on the junction of two semiconductors in a diode makes extra electrons available for one and extra holes for the other. These charges move apart and cause an electric current to flow through the wire connecting the two. This *photovoltaic effect* is the principle behind the solar cell. Spacecraft and space satellites in recent years have been equipped with large panels of solar cells that convert solar energy directly into electrical energy to operate the spacecraft (Fig. 11-6).

In our search for alternate forms of energy on earth, it

Fig. 11-6. **Large panels of solar cells on the NASA skylab space station convert solar energy directly into electrical energy.**

appears that solar cells may be one answer to the problem. If they can be made cheaply enough, large panels of solar cells may be used on the roofs of our homes and factories to generate electrical power from the energy of sunlight.

When radiant energy falls on some semiconductor metals, it also changes their ability to conduct electricity. Again, this effect is due to the fact that the energy knocks tightly-bound electrons up into an electron level where they can serve as current carriers. Such materials are called photo-conductors and are used in computers, radio, television, and other electronic equipment.

The effects of heat on semiconductors make possible the *thermoelectric generator.* This device is made by fastening an n-type and a p-type semiconductor to a connecting strap. When heat is applied to the strap, it makes that end of each semiconductor hotter than the opposite ends. Electrons and holes tend to move from the hot end to the cool end of their respective semiconductor. The pile-up of electrons and holes, as shown in Fig. 11-7, produces an electrical pressure, or voltage, that pushes an electric current through the connecting circuit. The thermoelectric generator thus converts heat energy directly into electrical energy. Such devices are most useful where only a small amount of electricity is needed, and where the low efficiency of these devices is not important. A heat source no larger than a kerosine lamp flame can generate enough power to operate a radio receiver. Such devices are used in remote areas of the world where no other source of electrical power is available.

The principle of the thermoelectric generator can be used

Fig. 11-7. **Semiconductors can be used as thermoelectric generators and to move heat from one point to another.**

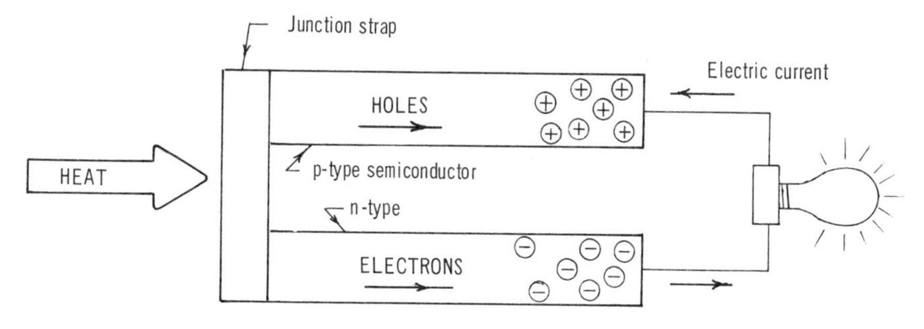

(a) Converting heat into electrical energy.

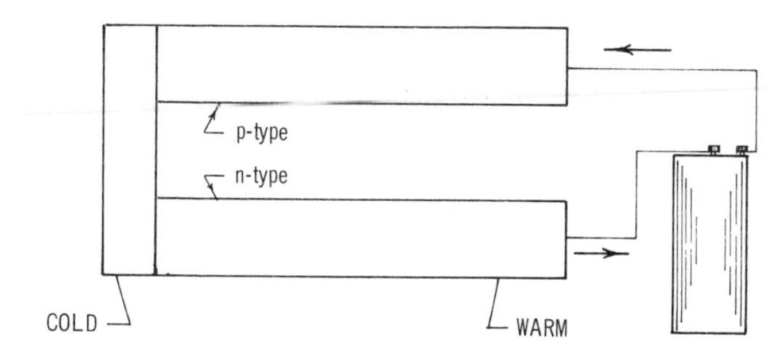

(b) Using electrical energy to pump heat.

in an opposite way; that is, to make an electric current move heat from one place to another. If an electric current is passed through a similar semiconductor device, the electrical field will move heat from one end to the other. The effect is so powerful that the cold end can be used to keep tiny specimens frozen while being observed under a microscope. Although such small-scale uses of this principle are the most common, household refrigerators have been built using semiconductor devices. Although such refrigerators are very simple, with no need for the pumps, blowers, and coils of a conventional refrigerator, they are not very efficient and therefore use a great deal of electrical power.

The ability of an element to conduct electricity, as we have seen in earlier chapters, depends not only upon the availability of free electrons, but also upon their ability to move through the crystals. In semiconductors, conductivity depends more upon availability of electrons than on other factors. As the temperature of a semiconductor increases, more electrons are given enough energy to jump to a level where they can act as conductors, and the conductivity increases.

In most ordinary metals, where there are always plenty of electrons available, the opposite is the case. Conductivity *decreases* as the metal is heated, because heat increases the vibrations of the atoms in the crystal. These vibrations interfere with the travel of the electrons and cause them to scatter, thus increasing resistance to the electric current. Conversely, as the metal is cooled, the vibrations of the atoms quiet down and allow freer passage of the electrons.

If the temperature of any material is reduced lower and lower, the vibrations of the atoms steadily decrease. At the lowest temperature that theoretically can be reached, called "absolute zero" which is –273°C (–460°F), all atomic vibrations presumably cease. At this temperature, it might be expected that the resistance of a metal to the flow of electricity would be zero. Unfortunately, absolute zero can never be reached, because it is impossible to remove all heat from a substance. However, experiments in 1911 by a Dutch physicist, H. K. Onnes (1853–1926), showed, to everyone's surprise, that when the temperature of a mercury wire was lowered to within about 4°C of absolute zero, it suddenly seemed to lose all resistance and become a perfect conductor of electricity.

This effect, known as *superconductivity*, is now known to occur with over two dozen metals, including the well-known metals aluminum, lead, tin, and mercury. Experiments have shown that the resistance does not simply become very small, but actually disappears completely and quite suddenly when the metal is lowered below its "critical temperature," which is usually between 1° and 10°C from absolute zero. In some experiments, an electric current has been started in a circular wire and found to be still flowing undiminished after one year. A number of interesting uses are possible for this phenomenon, since huge currents of electricity can be carried without loss, if the conductor can be cooled to the required low temperature. Such uses may include computer electronics and the generation of very strong magnetic fields.

The busy little electron scurrying through metals and semiconductors has brought us wonders that we sometimes take for granted. Radio, telephone, television, radar, and computers have all come into existence during the seventy-five years or so since the discovery of the electron. The latest triumphs include interplanetary travel and the return of photographs by television from the surface of the planet Mars. Who knows what will be next!

CHAPTER 12

RESOURCES, ENERGY, AND POLLUTION

The earth is a vast storehouse of energy, metal ores, and other resources. We take a variety of materials from the earth in enormous quantities. Many of these materials need little or no processing and are used almost as found. Water, for example, is often used just as it comes from the pump. Coal comes from the ground ready to burn. Petroleum needs only to be separated into various fuels and lubricants by simple distillation. (Other uses of petroleum may require much more complicated processing in refineries.) We use stone, both common and precious, with no alteration except cutting to shape and polishing. Many chemicals, such as salt, sulfur, and borax are mined and used with little processing.

Metal, on the other hand, is different from most other materials we take from the earth. Because most metal is tightly locked up in its ore, a great deal of effort and energy must be expended to set it free. The metals industry is a major industry in our society today. It is one of the largest users of energy and one of the largest producers of pollution.

The use of energy by the metals industry begins at the mine. Energy is used in the form of explosives to blast the tunnels and loosen the ore, and in the form of fuel for the machines that dig away the earth and shovel up the ore. More fuel is needed to power the trucks, trains, and conveyors that carry the ore to the processing plant. Still more energy is used at the plant to drive the conveyors, crushers, and other equipment that concentrate the ore.

Smelting the metal from the ore uses the largest amount of energy of any of the steps—energy in the form of coal (coke) for iron, oil for copper, and electricity for aluminum, to name only three examples. Each pound of aluminum requires the expenditure of energy equal to over six pounds of coal. For its annual production, the aluminum industry in the United States uses energy equal to more than 27 million tons of coal. Steel, on the other hand, requires only a little over one pound of coal for each pound of steel. Our enormous production of steel, however, makes the total annual use of energy in steel production equivalent to more than a quarter of a billion tons of coal—greater than that for any other metal.

To these numbers must be added the energy used in the factories that remelt basic metals to make alloys, and in the factories that cast, roll, draw, and cut metal into sheets, bars, tube, and wire. Still more fuel energy is needed to run the machines that cut, forge, stamp, machine, extrude, and weld metal into finished products for our use.

We know that simply burning this much fuel alone adds

considerably to atmospheric pollution. When we add the dust produced in mining and grinding the ore, the water used by all stages of metal production, and the particles and chemical fumes released in smelting the metal, we can see that the metals industry creates a large amount of pollution.

The metals that the industry produces are, of course, useful and vital in our society. If they are not properly handled, however, these same metals and their compounds or by-products can be serious health hazards, not only to the worker in the factories that produce them, but to people living nearby in the community. For example, pollutants that may be released in the manufacture of the metals mercury, lead, cadmium, and beryllium are toxic (poisonous) to living things. These poisons tend to persist in the environment —that is, they do not quickly break down into less harmful forms or substances.

Some metals can affect the central nervous system in humans and cause behavioral disorders. Lead, mercury, and manganese are particularly dangerous in this way. Compounds of selenium and antimony are liver poisons, while compounds of mercury and cadmium tend to attack the kidneys. Beryllium, cadmium, aluminum, tungsten carbide, and cobalt can harm the respiratory system if inhaled into the lungs in the form of dust or fumes. Some metals, such as nickel and the ores of iron and radioactive metals (such as uranium) can cause cancer.

The metal lead is a good example of one which is very useful but very dangerous. Compounds of lead were commonly used to make paints for many years. Lead paint flakes that have peeled from woodwork in homes have caused brain damage or death to small children who ate them. Lead paint has been chewed from toys with similar results. Although lead paints have been prohibited in the manufacture of toys for many years and are used little in household paints today, there are still a great many older homes where old flaking paint is a serious hazard to young children. Workers in the lead industry have been poisoned, too. The first signs of lead poisoning are fatigue, disturbance of sleep, intestinal colic, and constipation. Continued or severe exposure can cause brain damage and death. Tetraethyl lead [$Pb(C_2H_5)_4$] can cause similar disorders, leading to convulsions and death. Using this compound of lead in gasoline to increase its octane rating has put several hundred thousand tons of lead into the atmosphere each year from automobile exhausts. The use of it as an additive is being drastically reduced by the recent trend toward unleaded automobile fuels.

For many of these poisons, even a small amount is very dangerous. Two hundred parts of mercury in a *billion* parts of blood in the human body is an amount difficult to measure, but enough to cause noticeable symptoms of disease.

In the early days of the metals industry, workers in mines and mills were unprotected from the hazards around them. The dangers to their health were discovered the hard way—through sickness and death of workmen. Today elaborate precautions are taken to protect workers and to reduce health hazards to a minimum. A special branch of the medical profession, called *industrial medicine,* is engaged in researching health hazards and in improving the protection of workers from these hazards.

In recent years, we have become aware of the effects of poisons and pollutants that are released into the environment. For a very long time industrial wastes have been released into the air or dumped into the nearest stream. We have come to realize that many of these wastes do not "go away," even if they disappear from view. They remain in the environment until they are destroyed or rendered harmless by some natural process. In the meantime, they may harm or kill living creatures in the environment. Some poisons (such as mercury) can be picked up by water creatures, passed on to higher forms that eat these creatures, and finally move up the "food chain" to contaminate fish that are used for human food. In this way mercury that has been released into a river can end up poisoning living creatures —even people—hundreds of miles away and many years later.

The production of all of the three common metals—steel, aluminum, and copper—that we have followed through many chapters of this book has created serious environmental problems.

We might expect the environmental problems connected with iron and steel to be large and troublesome simply because the amount of these metals produced is so enormous. Annual world production of nearly a billion tons of steel begins with the mining of an even larger tonnage of ore. Digging up and carting away this much material lays waste to huge areas of the earth's surface (Fig. 12-1). When ore processing plants concentrate the ore (Fig. 12-2), mountains of waste material are discarded and must somehow be disposed of. One U.S. ore plant processing taconite, a hard-rock iron ore that is processed into ore-pellets for the blast furnace, produces 67,000 tons of waste material and 750 million gallons of waste water *per day*. This is enough waste material to fill 4 or 5 ocean-going freight ships and enough waste water to fill 50 or 60 large tanker ships. This material was all being dumped into one of the Great Lakes

Fig. 12-1. Mining metal ore can cause devastation to a large area of land. (*Kennecott Copper Corporation*)

Fig. 12-2. Ore processing plant. Note the waste dump near the top of the photograph. (*American Iron & Steel Institute*)

until the state and federal governments began to crack down. The waste water discharged included 60,000 pounds of dissolved solids each day, including such substances as ammonia, arsenic, cobalt, mercury, nitrates, and asbestos. Most of the substances are poisonous to some degree, and asbestos is known to cause cancer under some conditions. This plant is only one ore-processing facility out of the many in the United States today.

Smelting the iron requires more than a ton of coal for each ton of metal produced. Besides the pollutants produced by burning this much coal, huge quantities of dust, fumes, and waste water are produced in the smelting and steel-making processes (Fig. 12-3). These wastes, too, have been dumped into the atmosphere and into our rivers and lakes since these industries began. Only recently have state and local governments in the United States begun to require the industry to remove harmful substances before releasing waste

***Fig. 12-3.* Steel plant on Ohio River near Pittsburgh in 1961. (*Pennsylvania State Department of Health*)**

gases into the air and waste water into the world's water system.

Because aluminum is smelted by a process that uses electricity, we might expect that the electrolytic refining process might produce little or no pollution. Unfortunately, this is not the case. Some years ago, thousands of acres of pine trees in the Flathead National Forest and Glacier National Park began to show signs of disease. The United States Forest service was puzzled, because it could not find evidence of insect or fungus infection. Further investigation showed concentrations of fluorides (compounds of the poisonous gas fluorine) in the needles of these pines. The fluorides were causing branches to die back, causing needles to drop prematurely, reducing the rate of growth of the trees, and often killing the tree.

The source of the fluorides was found in an aluminum smelting plant in Montana, upwind from the damaged area. The fluorides came from the mineral cryolite, in which the alumina is dissolved for electrolysis (see Chapter 3). Cryolite is a double fluoride of aluminum. During the electrolysis process, fluorides were being released into the air and were spreading damage to the environment wherever the wind carried them. Damage to timber from this cause has been estimated at hundreds of millions of dollars.

In other areas farmers reported, and were able to prove, damage to livestock by fumes from aluminum smelters. Calves did not gain weight at the usual rate and some cattle died. Workers on the farms were annoyed by the fumes, and farmers found it difficult to sell hay and feed from fields that were known to be affected by smelter fumes.

The pollution caused by the manufacture of aluminum is not limited to fumes from the smelters. The smelting of aluminum consumes 8,000 watt-hours of electricity (enough to light 80 one hundred watt lamps for one hour) for each pound of aluminum produced. This means that over 6 pounds of fuel must be burned in a power plant to generate this much electricity. The environmental effects of mining and burning this much fuel must also be charged to the total cost of producing aluminum. (It might be argued that much of the aluminum in the United States is smelted by electrical energy obtained from water power, which is "nonpolluting." It may also be argued, however, that use of our nonpolluting electric resources for this purpose makes it necessary to produce an equivalent amount of electrical power by other means, such as coal, oil, and nuclear generator plants.)

The production of copper creates many similar problems, plus one that is peculiar to copper and some other metals: The most common ores of copper are sulfide ores; that is, compounds of copper and sulfur. When these ores are roasted in the smelting furnaces, sulfur compounds are released. If these compounds are allowed to escape into the air, they will later be washed back to the earth by rain as sulfuric acid. Sulfuric acid will rust metal, eat away the surface of stone statues and stone buildings, and is harmful to both plants and animals. Copper smelters have been noted for the devastation they bring to the area around them. The smelting of copper uses much less energy than aluminum, but nearly twice as much as steel.

New environmental regulations everywhere now require metal companies (and others) to greatly reduce the amount of pollutants they discharge into the world's water and air. Treatment systems are being used at the plant to clean the water used for various purposes of harmful pollutants before it is dumped back into the world's water cycle. Filtering and scrubbing systems are being used to clean smoke-stack discharges of harmful gases, dust, and fumes. These systems are often complicated and expensive, but in many cases can recover valuable substances or chemicals that can be sold or used for some purpose. But some of the material scrubbed from stack discharges is not useable and must be trucked away and buried. Many industrial areas of the world have greatly improved the quality of their air and water in recent years through pollution controls, but the job is sure to continue for a long time.

Because the cost of equipment to reduce pollution is so great, the argument is often heard that requiring such equipment will greatly increase the cost of the things we buy. Actually, the cost of things, as we have computed them in the past, has often been incorrect. The true cost of producing a pound of metal, for example, does not consist of only the direct costs that occur inside the doors of the factory. We must add in the cost of all the hidden effects, such as environmental damage, energy depletion, crop damage, health hazards, damages to buildings and bridges, etc., that may occur as a result of any of the processes. When these hidden costs are added up, it becomes clear that the cost of producing metal (or any other product) without pollution controls has not been as cheap as we had thought. By paying directly the costs of protecting the environment through pollution controls, we will in the end probably be spending no more for a product and will have a healthier and more pleasant environment in which to live.

Because we are using such large amounts of metals, we are rapidly depleting our ore resources. We are running out of the better-grade ores for many metals, and the industry is being forced to mine poorer ores which are often located in less accessible places. Mining these ores causes a greater amount of environmental devastation and requires more energy for transportation and processing.

We are also in danger of running out of ore for some important metals in the United States. Most of our supply of tin, aluminum, chromium, cobalt, nickel, manganese, and platinum already comes from outside the United States, and our dependence upon foreign sources is sure to increase in the future.

These problems of pollution, energy, and ore depletion would seem to suggest that we should be using metal wisely and conserving it wherever we can. This is not the case. The mountains of solid waste that threaten to bury the United States contain a great deal of metal—perhaps as much as 25 percent (Figure 12-4). Metal is one of the easiest materials to recycle—it can easily be melted down to make metal that is every bit as good as metal made from ore. Why then are we throwing metal away and then using great quantities of new ore and energy and creating more pollution to replace that metal? This question has no good answer—we must admit that what we are doing does not make sense.

Some metal is recycled, of course, but not nearly enough. The U.S. Environmental Protection Agency says that about 75 percent of the precious metals we consume is recycled metal. But among the less valuable metals, the amount recycled is much less. About one-fourth of our lead and copper comes from scrap, but only about one-fifth of the current production of iron and aluminum is made up of scrap. These numbers show that somewhere huge amounts of metals are being carted off to the dump and buried.

The EPA estimates that when we use scrap instead of new

ore to make iron and steel, we need only one-tenth as much new ore, about half as much water, and only one-fourth as much energy. Pollution of water and air would be reduced to one-fourth or less. In spite of these figures, the United States (at this writing) is exporting scrap iron to other countries and importing new iron ore!

One of the reasons for inadequate recycling of metal is that the metals industry is set up to handle production of metal from ore more cheaply than from scrap. Here again, if *all* the costs of production were considered, the answer would be different. When all of the hidden costs of producing metal are added in, including pollution, depletion of energy and ore resources, damage to health and to the environment, etc., it is clear that metal made from recycled scrap is much cheaper.

Fig. 12-4. **Discarding metal instead of recycling it wastes energy and metal resources and adds to the solid-waste problem. (*U.S. Environmental Protection Agency*)**

Recycling metal by remelting it may not always be the best way to conserve resources. There may be many cases where resources such as metals are being used for purposes that are inappropriate. The most obvious example of such uses is found in the container and packaging industry. There are many kinds of containers that should be made so that they can be *refilled* rather than either thrown away or recycled into new metal. Metal drink cans that are recycled, for example, must be collected, transported to the smelter, and melted down into new metal. The metal must be rolled into new sheet metal, treated with a rustproof coating, and manufactured into new cans. Then they must be shipped to the drink company, filled, and sent back to the store. This long process consumes several times as much energy as does the washing and refilling (in a local plant) of a reusable container, such as a glass bottle.

A refillable glass bottle can make many trips—from a dozen to perhaps as many as forty—between the bottler and the consumer before it will become so nicked and chipped that it must be melted down to make a new bottle. To be completely accurate, of course, we must include the cost of transporting the refillable bottles back to the drink factory, the cost of washing and sterilization, and the environmental cost of releasing the wash water and soap into the world's water system. Even when these costs are included, recycling by refilling is cheaper than recycling by remanufacture. (The cost of a gallon of soft drink in terms of energy required to manufacture the drink, the containers, and provide all transportation has been calculated at about four-tenths gallon of kerosine for canned drinks and one-eighth gallon of kerosine for drinks in refillable bottles.)

Recycling by remanufacture has an additional problem in

that much metal is thrown away rather than being offered for recycling. Only about one-seventh of the aluminum drink cans, for example, are (at this writing) being recycled—the rest are being thrown away and lost forever.

Because of the growing concern about conserving our resources of both energy and metal, more attention will probably be given in the future to using metal only for those purposes for which metal is really required. The day is probably not far off when we will recycle virtually all metal and probably most other materials. Perhaps it will some day be against the law to use metal for a frivolous purpose or to throw away a metal object instead of turning it in for recycling.

Metal has made life easier and better for the human race ever since it was first smelted some 6,000 years ago. Because metal is such a useful material, we mine enormous amounts of ore and smelt millions of tons of metal every year. Through carelessness and failure to consider the true cost of our actions, our production of metal is rapidly depleting the world's store of ores and energy, polluting water and air, and increasing solid wastes.

All of these problems can be solved by using metals more wisely, by protecting the environment from damage, and by recycling all metal. Fortunately, metal is different from many materials in that it can readily be recycled over and over again. Recycled metal will pour from the furnace as a bright, shiny stream of new metal, its atoms and electrons ready to go back to work, ready to do almost anything we ask them to do.

GLOSSARY

Technical language has been avoided or defined in everyday terms whenever possible in this book; however, it has been necessary to use many words that may be unfamiliar to the reader or which may have unfamiliar meanings when applied to metals. The words listed in this glossary are presented with simple explanations that pertain principally to their use in this book. More complete, formal definitions may be found in metallurgy textbooks and dictionaries.

ABSOLUTE ZERO The temperature at which all molecular activity is presumed to cease: –273°C, –460°F.

AGE-HARDENING Also called *precipitation-hardening.* The hardening of an alloy that results from the formation of tiny particles of an alloying element or a new mixture within a solid solution.

ALLOY A homogeneous mixture or solution of two or more metals, or of a metal and another element.

ALLOYING METAL A metal added to a larger amount of another metal in order to form an alloy.

ALUMINA An oxide of aluminum, Al_2O_3. Alumina is obtained from the ore bauxite and is used in the production of aluminum.

ALUMINUM OXIDE See *Alumina.*

AMORPHOUS Without definite form or structure; without a crystalline structure.

ANNEAL The process of heating a metal and cooling it slowly in order to soften it and reduce its brittleness.

ANODIZING Applying electrolytically a protective coating of metal oxide on a metal surface.

ANTHROPOLOGIST A scientist who studies the origin and cultural development of man.

ARGON An inert gaseous element found principally in the earth's atmosphere, which contains about 1 percent argon by weight.

ATOM The smallest unit of an element, consisting of a nucleus of protons and neutrons surrounded by a cloud of moving electrons; diameter approximately 0.00000001 centimeters.

ATOMIC NUMBER The number of protons in the nucleus of any element; also the number of planetary electrons in the atom.

ATOMIC WEIGHT The average weight of an atom of an element in atomic-weight units (awu), which unit has been chosen to give carbon an atomic weight of 12.

BAUXITE The principal ore of aluminum, containing alumina (Al_2O_3) and impurities.

BODY-CENTERED CUBIC A type of cubic crystal unit cell that has an atom at each corner and at the center of the cube.

BOND (chemical) The force which holds atoms or ions together in a molecule or crystal.

BRASS An alloy of copper and zinc with or without other metals in lesser amounts. See Figure 6-3.

BRAZING A process for joining two pieces of metal by melting between them a solder with a high melting temperature. See *soldering.*

BRITTLE Hard and rigid but likely to break if stretched or bent; it is the opposite of *plastic.*

BRONZE An alloy of copper, usually containing tin; more generally any alloy of copper not containing zinc. See Figure 6-3.

BRONZE AGE The historical period lying roughly between the discovery of bronze and the discovery of iron, during which tools and implements were made of bronze. See Figure 1-1.

CONDUCTOR (electrical) A material that will conduct an electric current.

CRUCIBLE PROCESS A steel-making process in which iron is heated with carbon in a sealed container (crucible) to carburize it into steel.

CRYSTAL An arrangement of atoms or molecules in a regular structure of identical unit cells. See Figure 5-2.

DEEP DRAWING The process of stretching a piece of metal, by pushing it through a die, into a deep-cup shape.

DIFFUSION The gradual mixing of the molecules of two or more substances as a result of heat-driven molecular motion.

DISORDERED (solid solution) A substitutional solid solution in which atoms of the alloying element and the parent metal are distributed at random through the crystal lattice. See *ordered.*

DRAWING The process of pulling a metal (e.g., a wire or tube) through a die to change its shape or size. See Figure 8-1.

DUCTILE Capable of being easily deformed or shaped, especially being stretched under tension, as in drawing a wire or tube.

ELASTIC DEFORMATION A temporary stretching, bending, or compressing of a metal; that is, the metal will return to its original shape when the force is removed. See *plastic deformation.*

ELECTROLYSIS The process of breaking down a compound into its components by passing an electric current through a solution. See Figure 3-4b.

ELECTRON One of the particles that make up the atom. It has a negative (–) electric charge. See *nucleus, proton, neutron.*

ELECTRON SHELLS Positions (energy levels) occupied by the electrons in orbit around the nucleus of an atom.

ELECTROPLATING To coat with a thin layer of metal by means of an electric current in an electrolyte.

ELEMENT A substance that cannot be separated into simpler substances by chemical means. There are ninety-two known natural elements in the universe.

EXTRUSION The process of shaping metal by forcing solid (usually hot) metal through a die. See Figure 8-1.

EUTECTIC An alloy having the lowest possible melting temperature for any alloy with those constituents.

EXTRACTIVE METALLURGY The science of taking metals from their ores and refining them.

FACE-CENTERED CUBIC A type of cubic crystal unit cell that has an atom at each corner and at the center of each face. See Figure 5-3.

FERROUS or ferric Pertaining to or containing iron.

FLUX A substance that helps to melt and remove the solid impurities (slag) in a smelting operation.

FORGING The process of forming metal by heating and hammering it into the desired shape.

FUME A discharge of smoke, vapor, or gas.

GALVANIZING The process of coating metal (usually iron or steel) with the metal zinc by dipping in molten zinc, by electrolytic action, or by spraying with molten zinc.

GRAIN A volume of metal having the same alignment of crystal structure throughout. The terms *grain* and *crystal* are generally used interchangeably.

GRAIN BOUNDARY (crystal boundary) The area or surface where one grain or crystal of metal is in contact with another. See Figure 5-6.

HARDNESS The resistance of a metal to bending, denting, or scratching.

HELIUM An inert gaseous element that is found on the sun and in natural gas on earth.

HEMATITE A blackish-red to brick-red mineral containing iron oxide, Fe_2O_3; one of the main ores of iron.

HEXAGONAL CLOSE-PACKED A crystal structure with a hexagonal (six-sided) base and top; an atom is located at each corner and three additional atoms inside the structure. See Figure 5-3.

HYDROGEN An active gaseous element which has the simplest atom of any element: one proton and one electron. Hydrogen is the most plentiful element in the universe.

INERT (element) An element that exhibits no chemical activity under ordinary conditions.

INSULATOR (electrical) A material that does not conduct electricity. See *conductor*.

INTERSTITIAL (solid solution) An alloy in which the atoms of the alloying element are located in the spaces between the atoms in the crystal lattice. See Figure 6-1.

ION An atom or molecule that has gained or lost one or more of its normal complement of electrons. An ion that has lost an electron has a positive charge; one that has gained an electron has a negative charge.

IONIC BOND The electrical attraction between a negatively charged ion and a positively charged ion that binds the two together. See Figure 4-5.

IRON AGE A period following the *Bronze Age* and beginning with the introduction of iron. See Figure 1-1.

KILN A high-temperature oven for baking or firing pottery. See Figure 1-4.

LATTICE The arrangements of atoms (or molecules) in a crystalline solid. See Figure 5-2.

MALLEABLE (metal) Capable of being permanently shaped or deformed under compression without breaking, as in rolling or hammering.

MELT To change from a solid to a liquid by the application of heat.

METALLOID A nonmetallic element that has some of the properties of a metal. Example: silicon.

METEORITE A stony or metallic object from space that survives the passage through the atmosphere and reaches the earth's surface. (*Meteoroid* refers to the object while it is in outer space, while *meteor* refers to the streak of light made by a meteoroid when it becomes incandescent in the earth's atmosphere, due to friction.)

MILLING Shaping a metal object by removing metal while moving it past a rotating cutter. See Figure 8-10.

MINERAL Generally, nearly any substance that occurs in nature that is neither animal nor vegetable. All metal ores and most rocks are considered minerals. Both petroleum and coal are sometimes defined as minerals, though both these substances are of organic origin.

MOLECULE The smallest particle of a substance (element or compound) that can exist in a free state.

MOLTEN Liquefied by heat.

NATIVE metal A metal which occurs in nature as a metal rather than as a compound or ore. Gold, silver, and copper are the most common native metals.

NEUTRON A particle that is found in the nuclei of all atoms except hydrogen. It has no electric charge and has approximately the same mass as a proton.

NEW STONE AGE The final period of the Stone Age, during which human beings domesticated animals, practiced farming, and made pottery, textiles, and a high grade of stone implements. Also called the *Neolithic Period.*

NOBLE element A chemically inactive or *inert* element.

OLD STONE AGE The early period of the Stone Age in which human beings made tools principally of bone, wood, and stone. Also called the *Paleolithic Period.*

ORE A mineral that is used as a source of a metal.

ORDERED (solid solution) A substitutional solid solution in which the atoms of the alloying element occupy certain regular positions in the crystal lattice of the parent metal. See Figure 6-1.

OXIDE A compound of an element and the element oxygen.

OXIDIZE To combine with oxygen.

PARENT METAL The metal comprising most of an alloy, to which smaller amounts of the alloying element are added. See *alloying element.*

PHYSICAL METALLURGY The science of converting a refined metal into a useful finished product. Includes alloying, heat treatment, forming, surface treatment, etc.

PLASTIC Capable of being permanently deformed without breaking. Opposite of brittle.

PLASTIC DEFORMATION A stretching, bending, or compressing of a metal that is permanent; that is, the metal will not return to its original shape after the force is removed. See *elastic deformation.*

POLLUTANT Any gases, particles, vapors, or chemicals that are not natural to air, soil, or water in both amount and kind.

PROTONS A particle that is found in the nuclei of all atoms. It has an electric charge (+) and a mass 1836 times that of an electron. See *neutron, electron.*

REDUCE (in metallurgy) To deoxodize, or break loose from oxygen; to smelt.

REDUCING AGENT A substance that chemically reduces another substance.

REFRACTORY Resistant to heat and high temperatures.

RIVET A metal pin, having a head at one end; used to fasten metal plates together by inserting the shank through holes in the plates and hammering the plain end into a new head.

ROLLING (in metallurgy) The process in which metal is passed between heavy metal rolls to decrease its thickness or to change its shape.

SLAG The mass of stony impurities left after a metal ore is smelted.

SLIP The movement or sliding of one part of a crystal with respect to another.

SMELTING The process of separating metal from its ore.

SOLDER An alloy of soft metals (usually tin and lead) with a low melting temperature; used to join metal parts together.

SOLDERING A process for joining two pieces of metal together by melting a low-melting-temperature solder between them.

SOLID SOLUTION A type of alloy in which atoms of the alloying metal occupy positions in the crystal lattice of the parent metal. See Figure 6-1.

SPECIFIC GRAVITY The ratio of the mass of a material to the mass of an equal volume of water. A specific gravity of four means a density four times that of water.

SPONGE Porous, finely divided copper or iron which was separated from impurities by low heat—so low that the ore does not completely melt.

STAMPING (in metallurgy) The process of forming or cutting metal by pressing it into a mold, form, or die.

STEEL An alloy of iron, containing about 0.08–2.0 percent carbon, with or without small quantities of other alloying metals.

STRENGTH The ability to resist deformation.

SUBSTITUTIONAL (solid solution) A type of alloy in which atoms of the alloying metal occupy positions in the crystal lattice of the parent metal. See Figure 6-1.

SULFIDE A compound composed of any element and the element sulfur.

TEMPERING The process of reducing the brittleness of a metal by heating it to a temperature well below its melting point and cooling it slowly.

TENSILE STRENGTH The force that a metal can withstand in tension (stretching) without breaking, usually expressed in pounds per square inch.

TOUGHNESS The ability of a metal to withstand blows and sudden loads without breaking.

TOXIC Harmful, poisonous.

TRANSMUTATION The conversion of one element into another, in particular, base metal (lead) into precious metal (gold). (The unaccomplished goal of alchemy.)

UNIT CELL The smallest geometric arrangement of atoms that represents the entire crystal. The unit cell is repeated over and over to make up a crystal lattice. See Figure 5-3.

WELDING The joining of metal parts by applying heat to melt adjoining surfaces; also by hammering, or compressing, especially after heating to red-hot condition.

WORK-HARDENING The hardening that results from cold-working a metal.

YIELD STRENGTH The force that a metal can withstand without undergoing a permanent change of shape, usually expressed in pounds per square inch.

SUGGESTED FURTHER READINGS

Because few books seem to have been written for the layman on the topic of metals, no list of such books can be offered. Students who wish to pursue specific topics to greater depth will find excellent material in good encyclopedias in articles on specific metals, metal-working processes, mining, metallurgy, etc., and under such related topics as crystal structure, nuclear physics, semi-conductors, etc.

Other sources of information include high-school and college texts on science, physics and chemistry, and college texts on metallurgy. Although college texts tend to be highly technical and assume considerable background knowledge, examining a number of them will often turn up a treatment of a particular topic that will be useful to the layman.

INDEX

ABOUT THE AUTHOR

NORMAN F. SMITH has been interested in science for as long as he can remember. Pestered endlessly with questions about meteorites and stars and how things worked, his father put him into science courses in high school and steered him into engineering in college. After graduating with distinction from Purdue University in Mechanical Engineering, Mr. Smith worked for NACA (the predecessor of NASA) in wind tunnel research, helping to solve some of the high-speed aerodynamic problems that plagued military airplanes during WW II. After the war he moved into supersonic research, working on the aerodynamics of supersonic bombers and transports long before such airplanes were a practical reality. During these research years, he authored more than a dozen technical reports published by NACA/NASA.

After the war, Mr. Smith fulfilled an old dream by learning to fly. He is now an active instrument-rated pilot.

When the American space program began, Mr. Smith moved into the NASA Space Task Group which later became the Manned Spacecraft Center in Houston, Texas. While working in a number of engineering and administrative positions during projects Mercury, Gemini, and Apollo, he also represented NASA at educational functions such as conventions, lectures, and science fairs, and became interested in science teaching and science writing. His first two books, quite naturally on the topics of space and aeronautics, were published in 1970 and 1972 while he was still employed by NASA. "They were written," Mr. Smith says, "in an attempt to get better translations of the science behind these two topics into the hands of science teachers, students, and laymen."

Mr. Smith took early retirement near the end of the Apollo program to devote full time to writing children's books, filmstrips, movies, and magazine articles, all on his favorite topic: science. *The Inside Story of Metal* is his eighth book for young people. Mr. Smith and his wife live in the Champlain Islands of Vermont, where they cruise in a boat large enough not only to have a nook for writing, but also, he says hopefully, large enough to sail anywhere in the world.